SpringerBriefs in Electrical and Computer Engineering

Control, Automation and Robotics

Series Editors

Tamer Başar, Coordinated Science Laboratory, University of Illinois at Urbana-Champaign, Urbana, IL, USA

Miroslav Krstic, La Jolla, CA, USA

SpringerBriefs in Control, Automation and Robotics presents concise summaries of theoretical research and practical applications. Featuring compact, authored volumes of 50 to 125 pages, the series covers a range of research, report and instructional content. Typical topics might include:

- a timely report of state-of-the art analytical techniques;
- a bridge between new research results published in journal articles and a contextual literature review;
- a novel development in control theory or state-of-the-art development in robotics;
- an in-depth case study or application example;
- a presentation of core concepts that students must understand in order to make independent contributions; or
- a summation/expansion of material presented at a recent workshop, symposium or keynote address.

SpringerBriefs in Control, Automation and Robotics allows authors to present their ideas and readers to absorb them with minimal time investment, and are published as part of Springer's e-Book collection, with millions of users worldwide. In addition, Briefs are available for individual print and electronic purchase.

Springer Briefs in a nutshell

- 50–125 published pages, including all tables, figures, and references;
- softcover binding;
- publication within 9–12 weeks after acceptance of complete manuscript;
- copyright is retained by author;
- authored titles only—no contributed titles; and
- versions in print, eBook, and MyCopy.

Indexed by Engineering Index.

Publishing Ethics: Researchers should conduct their research from research proposal to publication in line with best practices and codes of conduct of relevant professional bodies and/or national and international regulatory bodies. For more details on individual ethics matters please see: https://www.springer.com/gp/authors-editors/journal-author/journal-author-helpdesk/publishing-ethics/14214

More information about this series at http://www.springer.com/series/10198

Juntao Chen · Quanyan Zhu

A Game- and Decision-Theoretic Approach to Resilient Interdependent Network Analysis and Design

Springer

Juntao Chen
Department of Electrical and Computer
Engineering, Tandon School of Engineering
New York University
Brooklyn, NY, USA

Quanyan Zhu
Department of Electrical and Computer
Engineering, Tandon School of Engineering
New York University
Brooklyn, NY, USA

ISSN 2191-8112   ISSN 2191-8120   (electronic)
SpringerBriefs in Electrical and Computer Engineering
ISSN 2192-6786   ISSN 2192-6794   (electronic)
SpringerBriefs in Control, Automation and Robotics
ISBN 978-3-030-23443-0   ISBN 978-3-030-23444-7   (eBook)
https://doi.org/10.1007/978-3-030-23444-7

Mathematics Subject Classification (2010): 91A80, 91A44, 90B10

© The Author(s), under exclusive license to Springer Nature Switzerland AG 2020
This work is subject to copyright. All rights are solely and exclusively licensed by the Publisher, whether the whole or part of the material is concerned, specifically the rights of translation, reprinting, reuse of illustrations, recitation, broadcasting, reproduction on microfilms or in any other physical way, and transmission or information storage and retrieval, electronic adaptation, computer software, or by similar or dissimilar methodology now known or hereafter developed.
The use of general descriptive names, registered names, trademarks, service marks, etc. in this publication does not imply, even in the absence of a specific statement, that such names are exempt from the relevant protective laws and regulations and therefore free for general use.
The publisher, the authors and the editors are safe to assume that the advice and information in this book are believed to be true and accurate at the date of publication. Neither the publisher nor the authors or the editors give a warranty, expressed or implied, with respect to the material contained herein or for any errors or omissions that may have been made. The publisher remains neutral with regard to jurisdictional claims in published maps and institutional affiliations.

This Springer imprint is published by the registered company Springer Nature Switzerland AG
The registered company address is: Gewerbestrasse 11, 6330 Cham, Switzerland

To our families
Juntao Chen and Quanyan Zhu

Preface

This book provides an overview of game and decision theoretic methods for designing resilient and interdependent networks. The book aims to unite game theory with network science to lay a system-theoretic foundation for understanding the resiliency of interdependent and heterogeneous network systems. One focused application area of the book is the critical infrastructure systems. Infrastructure networks such as electric power, water, transportation, and communications are increasingly interconnected with the integration of Internet of Things devices. A single-point shutdown of a generator in the electric power either due to cyber and physical attacks or natural failures can propagate to other infrastructures and creates an enormous social and economic impact. Therefore, secure and resilient design of interdependent critical infrastructure is critical. To achieve this goal, it requires multidisciplinary research in this area that crosscuts computer science, engineering, public policy, social sciences, and mathematics. The book summarizes recent research findings into three parts including resilient meta-network modeling and analysis, control of interdependent epidemics spreading over large-scale complex networks, and applications to critical infrastructures such as Internet of battlefield things. Each chapter includes a section on background, which does not require the readers of this book to have advanced knowledge in game and decision theory and network science.

The book is self-contained and can be adopted as a textbook or supplementary reference book for courses on network science, systems and control theory, and infrastructures. The book will be also useful for practitioners or industrial researchers across multiple disciplines including engineering, public policy, and computer science who look for new approaches to assess and mitigate risks in their systems and enhance their network resilience.

The authors would like to thank fruitful discussions and collaborations with Corrine Touati (INRIA, France), Rui Zhang (NYU), and other research members in NYU Tandon LARX. The authors would also like to acknowledge support from NSF and DHS.

Brooklyn, NY, USA
May 2019

Juntao Chen
Quanyan Zhu

Contents

Chapter 1
Introduction

1.1 Motivation and Introduction

Our world is increasingly connected due to the adoption of smart devices and Internet of Things (IoT). Not only does the connectivity of the network itself grows but also networks are interconnected with other networks which create interdependent networks. For example, the power networks are nowadays integrated with communication networks with the advances of the smart grid technologies. Transportation networks are connected with social networks through on-demand transport systems. The deeply interconnected cyber-physical-social networks create new challenges for improving the resiliency at different scales against all hazards from nature, terrorism, and deliberate cyber attacks.

The first challenge of designing resilient interdependent networks comes from the lack of system framework that captures heterogeneous network components. The existing models in literature are mostly designed for a single-layer network containing a number of agents. In this book, we propose a *network-of-networks* framework that jointly considers the interactions within a network itself and across different layers of networks. This framework facilitates the analysis of network operators' strategies whose objectives and actions are coupled due to the inherent network interdependencies. The network-of-networks modeling offers a holistic view of the separate components by leveraging which we can analyze the *system-of-systems* performance of the global network.

The second challenge for designing resilient interdependent network is the *uncoordinated* nature between system designers. This characteristic has been observed in a number of scenarios. For example, the power system and transportation system operators determine their operational policies separately with a goal in improving their own revenue even though these two networks are coupled. This decision-making pattern is different from single-layer network where the designer maximizes the global system utility. To address this distinct challenge in interdependent networks, we establish a *game-theoretic* framework to capture the decentralized nature of decision-making.

© The Author(s), under exclusive license to Springer Nature Switzerland AG 2020
J. Chen and Q. Zhu, *A Game- and Decision-Theoretic Approach to Resilient
Interdependent Network Analysis and Design*, SpringerBriefs in Control,
Automation and Robotics, https://doi.org/10.1007/978-3-030-23444-7_1

The interactions between different networks can be viewed as a noncooperative game in which each network optimizes its own objective. The resulting equilibrium solution predicts the outcome of such strategic interactions which further provides analytical basis for designing mechanisms to build interdependencies that yield desirable network-of-networks at equilibrium.

Human and social networks are another important class of networks of which optimal and secure control design is critical. Similar to computer networks, one feature of human or social networks is its large number of agents. Due to the enormous scale of the network, designing explicit strategy for each agent becomes prohibitive or even impossible. To address this challenge, we need to shift the focus from fine-grained modeling to approximate modeling of the complex network while preserving the interdependencies between agents. Therefore, we establish a mean-field approximation framework by classifying the nodes in the network according to their degrees. This convenient modeling facilitates the analysis and design of control policies of interdependent epidemics over complex networks. The application scenarios include spreading control of viruses and ransomware on the Internet, and diseases such as Ebola in the human society.

Another critical factor needs to consider in resilient network design is the implementation complexity of strategies. A system with agile resilience requires an efficient recovery policy which can be computed and implemented easily. In the meta-network resilience game, we transform the originally formulated game problem into semidefinite programs which can be solved efficiently. The interdependent mobile autonomous system is resilient if the control policy is situationally aware. We design online control algorithms to achieve this goal to optimize the network resilience. In addition, the devised algorithm for constructing optimal secure interdependent infrastructure network scales well with linear complexity in the size of networks.

1.2 Overview of the Book

The rest of the book is organized as follows. In Chap. 2, we will briefly present the basics of game theory and network science which are theoretical foundations of the entire book. In Chap. 3, we will establish static and dynamic interdependent network resilience game in which each designer determines the strategy for his own subnetwork. We further devise decentralized and computationally efficient policies for the system designers in optimizing their aligned goals on network-of-networks performance. In Chap. 4, we expand the scope from finite networks which is a focus of Chap. 3 to complex networks. This large-scale network modeling is able to capture the system with a large number of population, e.g., social and human networks and computer networks. Based on the established model, we investigate the optimal control of interdependent epidemics spreading over complex networks. The obtained results provide guidelines for network operators in controlling interdependent diseases and viruses by considering tradeoffs between epidemics severity and applied effort costs. Knowing that critical infrastructures could be disconnected due to cyber and physical attacks which downgrade the network efficiency, the designer should take security considerations into account when designing the network at the beginning. Chapter 5 aims to address this challenge by developing a systematic approach for designing secure multilayer networks. Since different layers face various levels of cyber threats, the system operator needs to design the multi-layer network with heterogeneous security requirements for each subnetwork. Under the limited budget constraint, we characterize the optimal design solutions and propose an algorithm to construct the optimal network, and illustrate the design principles through multi-layer Internet of battlefield things (IoBT). Finally, we conclude the book and discuss future works in Chap. 6.

Chapter 2
Background of Game Theory and Network Science

2.1 Introduction to Game Theory

Game theory is widely adopted in modeling and analyzing strategic interactions between a number of independent agents (also called *players*) [1, 2]. A game $\mathcal{G}$ can be generally defined by a tuple $\mathcal{G} := \{\mathcal{N}, (\mathcal{A}_i)_{i\in\mathcal{N}}, (U_i)_{i\in\mathcal{N}}\}$, where $\mathcal{N}$ is the set of players, $\mathcal{A}_i$ is the action set of player i, and U_i is the utility function of player i. Specifically, we consider an N-player game, where $\mathcal{N} := \{1, 2, \ldots, N\}$. The decision variable of player $i \in \mathcal{N}$ is denoted by $a_i \in \mathcal{A}_i$. Note that the action set can be finite (infinite) such that players have a finite (infinite) number of possible actions. For convenience, we denote the action of all N players as $a := (a_1, a_2, \ldots, a_N)$. In addition, denote by $\mathcal{A}$ the Cartesian product in the form of $\mathcal{A}_1 \times \mathcal{A}_2 \times \cdots \times \mathcal{A}_N$.

The utility function of player i can be explicitly written as $U_i(a_i, a_{-i}) : \mathcal{A} \to \mathbb{R}$, where a_{-i} denotes the action profile of all players except the ith one. Furthermore, we denote by $\Omega \subset \mathcal{A}$ the feasible action set of all players after capturing the possibly coupled constraints. Thus, a feasible a needs to satisfy $a \in \Omega$. We further denote by Ω_i the constrained action set of player i. A game is finite, in contrast to infinite games (or continuous-kernel games), if both the action set and the number of players is finite.

2.1.1 Finite Nash Games

In finite games, each player chooses actions from the action set, either in pure strategy or mixed strategy sense, to maximize its utility. A pure strategy indicates that the player chooses a single action with certainty. In comparison, a mixed strategy is represented by a probability distribution over the action set which specifies the probability of taking each action.

To characterize the strategic behaviors of players, we adopt Nash equilibrium (NE) as the solution concept which is defined as follows.

© The Author(s), under exclusive license to Springer Nature Switzerland AG 2020
J. Chen and Q. Zhu, *A Game- and Decision-Theoretic Approach to Resilient Interdependent Network Analysis and Design*, SpringerBriefs in Control, Automation and Robotics, https://doi.org/10.1007/978-3-030-23444-7_2

Definition 2.1 (*Nash Equilibrium*) An N-tuple action profile $a^* \in \Omega$ constitutes a Nash equilibrium (NE) of game $\mathcal{G}$ if, for all $i \in \mathcal{N}$,

$$U_i(a_i^*, a_{-i}^*) \geq U_i(a_i, a_{-i}^*), \ \forall a_i \in \mathcal{A}_i, \ \text{such that} \ (a_i, a_{-i}^*) \in \Omega. \tag{2.1}$$

Since a finite N-player game may not have an NE in pure strategy, the solution concept can be extended to mixed strategy NE. The mixed strategy of player i is denoted by p_i which assigns the probability of taking each action. In search of a mixed strategy equilibrium, U_i is replaced by its expected value taken with respect to the mixed strategy choices of the players, which we denote for player i by $L_i(p_1, \ldots, p_N)$. Denote $\mathcal{P}_i$ by the set of all probability distributions on $\mathcal{A}_i$. Then, the definition of mixed strategy NE is given as follows.

Definition 2.2 (*Mixed Strategy Nash Equilibrium*) An N-tuple action profile $(p_1^*, \ldots, p_N^*)$ constitutes a mixed strategy NE of game $\mathcal{G}$ if, for all $i \in \mathcal{N}$,

$$L_i(p_i^*, p_{-i}^*) \geq L_i(p_i, p_{-i}^*), \ \forall p_i \in \mathcal{P}_i. \tag{2.2}$$

The existence of NE in finite Nash games is presented below whose proof can be found in [3].

Theorem 2.1 *Every finite N-player nonzero-sum game has a Nash equilibrium in mixed strategies.*

2.1.2 Infinite Nash Games

In an N-player infinite game, the action set $\mathcal{A}_i$ is a finite-dimensional space instead of a finitely countable set, $\forall i \in \mathcal{N}$. The utility function U_i is a function on the finite-dimensional product space $\mathcal{A}$. The definition of NE strategies of infinite games is the same as the ones in Definitions 2.1 and 2.2 with only slightly differences in the redefined action sets.

The results of existence of NE strategy are summarized in the following Theorems 2.2 and 2.3 which can be found in [4].

Theorem 2.2 *In the N-player nonzero-sum infinite game, if the constrained action set Ω_i for player i is a closed and bounded subset of a finite-dimensional Euclidean space, and $U_i(a_i, a_{-i})$ is continuous for each $i \in \mathcal{N}$, then there exists an NE in mixed strategies.*

Theorem 2.3 *In the N-player nonzero-sum infinite game, if the constrained set Ω is a closed, bounded, and convex subset of a finite-dimensional Euclidean space, and $U_i(a_i, a_{-i})$ is strictly concave in a_i for each a_{-i} and each $i \in \mathcal{N}$, then there exists an NE in pure strategies.*

2.1.3 Stackelberg Games

The Nash equilibrium solution concept provides a noncooperative equilibrium solution for nonzero-sum games when the roles of the players are symmetric. However, when one of the players has the ability to enforce his strategy on the others, one needs to introduce a hierarchical equilibrium solution concept. Following the terminology in [5], the player who dominates the game is called the *leader*, and the others reacting to the leader's strategy are called the *followers*. For the sake of clarity in exposition, we focus on presenting a two-person Stackelberg games in this section, i.e., one leader and one follower. A number of extensions of the Stackelberg solution concept to N-person static games with different levels of hierarchy can be found in [5].

Before presenting the solution concept for Stackelberg game, we introduce the following definition.

Definition 2.3 (*Best Response*) In a two-person static game, the set $BR_2(a_1) \subset \Omega_2$ defined for each $a_1 \in \Omega_1$ by

$$BR_2(a_1) \subset \Omega_2 = \{\zeta \in \Omega_2 : U_2(a_1, \zeta) \geq U_2(a_1, a_2), \ \forall a_2 \in \Omega_2\} \qquad (2.3)$$

is the best response set of player 2 to the strategy $a_1 \in \Omega_1$ of player 1.

Based on the best response definition, we define Stackelberg equilibrium solution concept as follows.

Definition 2.4 (*Stackelberg Equilibrium*) In a two-person game with player 1 as the leader, a strategy $a_1^* \in \Omega_1$ is called a Stackelberg equilibrium strategy for the leader if

$$\min_{a_2 \in BR_2(a_1^*)} U_1(a_1^*, a_2) = \max_{a_1 \in \Omega_1} \min_{a_2 \in BR_2(a_1)} U_1(a_1, a_2). \qquad (2.4)$$

Remark 2.1 If $BR_2(a_1)$ is a singleton for each $a_1 \in \Omega_1$, i.e., the best response function of player 2 is described completely by a reaction curve $l_2 : \Omega_1 \to \Omega_2$, then (2.4) can be replaced by

$$U_1(a_1^*, l_2(a_1^*)) = \max_{a_1 \in \Omega_1} U_1(a_1, l_2(a_1)). \qquad (2.5)$$

The existence of Stackelberg equilibrium is summarized below. More discussions on the properties on Stackelberg equilibrium can be found in [5].

Theorem 2.4 *The following statements hold:*

(1) Every two-person finite game admits a Stackelberg strategy for the leader, and the follower's strategy is characterized by the best response.

(2) In two-person infinite games, let Ω_1 and Ω_2 be compact subsets, and U_i be continuous on $\Omega_1 \times \Omega_2$, $i = 1, 2$. Let there exist a finite family of continuous

mappings $l_i : \Omega_1 \to \Omega_2$, *with* $i \in I := \{1, \ldots, M\}$, *such that* $BR_2(a_1) = \{a_2 \in \Omega_2 : a_2 = l_i(a_1), i \in I\}$. *Then, the two-person nonzero-sum infinite game admits a Stackelberg equilibrium strategy.*

The readers interested in a complete introduction of game theory can refer to [1, 5] for more details.

2.2 Basics of Network Science

2.2.1 Modeling of Networks

An undirected graph G is defined by a pair of sets (V, E), where V is a non-empty countable set of elements, called *nodes* or *vertices*, and E is a set of unordered pairs of different nodes, called *edges* or *links*. The link (i, j) joins the nodes i and j. The total number of nodes in the graph is equal to the cardinality of the set V denoted by $|V|$ which is also referred as the size of the graph G. The cardinality of the set E is equal to the number of edges. Note that in a graph with n nodes, the maximum number of links is equal to $\frac{n(n-1)}{2}$. When all pairs of nodes are connected, then G is called a complete graph.

Suppose that $G(V, E)$ consists of n nodes and interconnected by m links. To represent an undirected graph $G(V, E)$, *adjacency matrix* is usually adopted. Denote the adjacency matrix of G by $\mathbf{A} \in \mathbb{R}^{n \times n}$. The element of $\mathbf{A}$ is denoted by a_{ij} taking values as follows:

$$a_{ij} = \begin{cases} w_{ij}, \text{nodes } i \text{ and } j \text{ are connected}; \\ 0, \text{nodes } i \text{ and } j \text{ are not connected}; \end{cases} \tag{2.6}$$

where $w_{ij} \in \mathbb{R}_+$ is the link weight. If G is an unweighted graph where links are homogeneous, then $w_{ij} = 1$ if link $(i, j) \in E$, and otherwise $w_{ij} = 0$. When G is a weighted graph where each link (i, j) is associated with a weight w_{ij} representing the intensity of its connection, then the entry in $\mathbf{A}$ becomes $a_{ij} = w_{ij}$ if link $(i, j) \in E$, and otherwise $a_{ij} = 0$.

There are a number of metrics to quantify the performance of graph for different purposes, including the node degree, nearest neighbors, reachability, shortest path, and diameter. We focus on a metric called *algebraic connectivity* [6] which is an indicator of how well a graph is connected. Algebraic connectivity is based on the Laplacian matrix of a graph. Consider a graph consists of n nodes and m links. For a link l that connects nodes i and j where the link weight equaling to w_{ij}, we define two n-dimensional vectors $\mathbf{a}_l$ and $\mathbf{b}_l$, where $\mathbf{a}_l(i) = 1$, $\mathbf{a}_l(j) = -1$, $\mathbf{b}_l(i) = w_{ij}$, $\mathbf{b}_l(j) = -w_{ij}$, and all other entries 0. When G is unweighted, $w_{ij} = 1$. Then, the Laplacian matrix $\mathbf{L}$ of network G can be expressed as

$$\mathbf{L} = \sum_{l=1}^{m} \mathbf{a}_l \mathbf{b}_l^T, \tag{2.7}$$

where "T" denotes the matrix transpose operator. Intuitively, the ith diagonal entry $\mathbf{L}_{ii}$ in the Laplacian matrix is equal to the degree of node i, i.e., $\mathbf{L}_{ii} = \sum_{j \in \mathcal{N}_i} w_{ij}$, $\forall i \in V$, where $\mathcal{N}_i$ denotes the set of nodes that connects with node i. In addition, $\mathbf{L}_{ij} = -w_{ij}$, $\forall i \neq j \in V$, if nodes i and j are connected, and otherwise is 0. Additionally, Laplacian matrix is positive semidefinite, and $\mathbf{L}\mathbf{1} = 0$, where $\mathbf{1}$ is an n-dimensional vector with all one entries. Thus, by ordering the eigenvalues of $\mathbf{L}$ in an increased way, we obtain

$$0 = \lambda_1 \leq \lambda_2 \leq \cdots \leq \lambda_n, \tag{2.8}$$

where the smallest eigenvalue $\lambda_1(\mathbf{L}) = 0$, and $\lambda_2(\mathbf{L})$ is called algebraic connectivity (or Fiedler value) of G [6]. Further, $\lambda_2(\mathbf{L}) = 0$ when G is not connected. For a graph with Laplacian $\mathbf{L}$, the algebraic connectivity $\lambda_2(\mathbf{L})$ can be computed from the Courant–Fisher theorem [7] as follows:

$$\lambda_2(\mathbf{L}) = \min\{z^T \mathbf{L} z | z \in \mathbf{1}^\perp, ||z||_2 = 1\}, \tag{2.9}$$

where $|| \cdot ||_2$ denotes the standard L_2 norm.

The readers interested in more detailed and formal discussions on graph theory can refer to [8, 9].

2.2.2 Modeling of Network-of-Networks

To facilitate the analysis and design of resilient interdependent networks, we need to establish a model for network-of-networks. We consider two interdependent networks $G_1(V_1, E_1)$ and $G_2(V_2, E_2)$, where networks 1 and 2 are represented by the graphs G_i, $i = 1, 2$, respectively. Network i, for $i \in \{1, 2\}$, is composed of $n_i = |V_i|$ nodes and $m_i = |E_i|$ links. The set of links denoted by E_i are called the *inter-links* of individual network i. The two networks can also be connected using *intra-links* which creates the interdependencies between two networks. Let E_{12} be the set of m_{12} intra-links between G_1 and G_2, with $m_{12} = |E_{12}|$. Hence, the global network can be represented by the combined graph $G = (V_1 \cup V_2, E_1 \cup E_2 \cup E_{12})$. Note that we also use E_{21} to denote the set of intra-links in G for convenience later, and thus $E_{21} = E_{12}$. Let $n = n_1 + n_2$ and $m = m_1 + m_2 + m_{12}$. The adjacency matrix $\mathbf{A} \in \mathbb{R}^{n \times n}$ of the global networks G has the entry a_{ij} defined in (2.6). In the following Chap. 3, $w_{ij} = 1$ in the static network resilience game, while the value of w_{ij} in the dynamic resilience game counterpart depends on the distance between nodes i and j.

Let $\mathbf{A}_1 \in \mathbb{R}^{n_1 \times n_1}$ and $\mathbf{A}_2 \in \mathbb{R}^{n_2 \times n_2}$ be the adjacency matrices of G_1 and G_2. When these two networks are disconnected, $\mathbf{A}$ takes the following form

$$A = \begin{bmatrix} A_1 & 0_{n_1 \times n_2} \\ 0_{n_2 \times n_1} & A_2 \end{bmatrix},$$

where $0_{n_1 \times n_2}$ is an $n_1 \times n_2$-dimensional matrix with all zero entries. When $E_{12} \neq \emptyset$, the adjacency matrix of the network G becomes

$$A = \begin{bmatrix} A_1 & B_{12} \\ B_{12}^T & A_2 \end{bmatrix},$$

where $B_{12} \in \mathbb{R}^{n_1 \times n_2}$ is an off-diagonal block matrix used to capture the effect of intra-links between networks.

The Laplacian matrix L can be rewritten as adjacent matrices A_1 and A_2. Let $D_1 \in \mathbb{R}^{n_1 \times n_1}$ and $D_2 \in \mathbb{R}^{n_2 \times n_2}$ be two diagonal matrices associated with network 1 and 2, respectively, which are defined as follows:

$$\begin{cases} (D_1)_{ii} = \sum_j (B_{12})_{ij}, \\ (D_2)_{ii} = \sum_j (B_{12}^T)_{ij}. \end{cases}$$

Then, by using $L = D - A$, the Laplacian matrix of G is

$$L = \begin{bmatrix} L_1 + D_1 & -B_{12} \\ -B_{12}^T & L_2 + D_2 \end{bmatrix}, \tag{2.10}$$

where $L_i = D_i - A_i$, $i = 1, 2$, are Laplacians associated with G_1 and G_2, respectively.

Remark 2.2 The above modeled two-layer interdependent networks can be easily extended to multilayer cases.

2.3 Notation Conventions

The conventions in the rest of the book are summarized as follows. The bold symbol refers to either a vector or a matrix. $\mathbb{R}$ and $\mathbb{Z}$ refer to the real numbers and real integers, respectively. $|\cdot|$ denotes the cardinality of a set. $||\cdot||_2$ denotes the standard L_2 norm. λ_2 represents the algebraic connectivity of a network. Superscript T denotes the transpose of a vector or a matrix. I and 1 represent an identity matrix and a vector with all one entries, respectively. L denotes the Laplacian matrix of a network. P_i and G_i stand for player i and network i, respectively.

References

1. Fudenberg D, Tirole J (1991) Game theory. MIT press, Cambridge
2. Owen G (2013) Game theory, 4th edn. Academic, New York
3. Nash JF et al (1950) Equilibrium points in n-person games. Proc Natl Acad Sci 36(1):48–49
4. Glicksberg IL (1952) A further generalization of the kakutani fixed point theorem, with application to nash equilibrium points. Proc Am Math Soc 3(1):170–174
5. Başar T, Olsder GJ (1999) Dynamic noncooperative game theory, 2nd edn. Classics in applied mathematics. SIAM, Philadelphia
6. Fiedler M (1973) Algebraic connectivity of graphs. Czechoslov Math J 23(2):298–305
7. Horn RA, Johnson CR (2012) Matrix analysis. Cambridge University Press, Cambridge
8. West DB (1996) Introduction to graph theory, vol 2. Prentice hall, Upper Saddle River
9. Bollobás B (2013) Modern graph theory, vol 184. Springer, Berlin

Chapter 3
Meta-Network Modeling and Resilience Analysis

3.1 Static Network Resilience Game

In this section, we investigate the static network resilience game. Its dynamic network resilience game counterpart will be studied in Sect. 3.2.

The recent advances in information and communications technologies (ICTs), such as 5G wireless networks and the Internet of Things (IoTs), have created a growing amount of connectivity between systems. The interconnection between network systems creates a network-of-networks in which the interdependencies play an important role in understanding their emerging functions and performances. The connectivity between the interdependent networks is critical for the robustness and resilience of the entire system. A higher connectivity allows faster information exchange and quick emergency response to natural or man-made disasters, and hence increases the ex post resilience.

One important type of network-of-networks comes from the interaction between different modern critical infrastructures [1–3]. For example, Fig. 3.1 illustrates a two-layer interdependent network including the communication and power networks. Both network designers are making decisions on the network configuration for better connectivity. One the one hand, routers in the communication network can set up wireless communication links with other routers and substations for information exchange. On the other hand, substations in the power network can form connections with other substations through power line communication (PLC) or wireless channel with routers for communication. Though the goals of the two network operators are aligned, they do not coordinate to achieve their objectives which will lead to potential inefficiency. Therefore, one challenge for the interdependent critical infrastructures network design is the decentralized nature of decision making.

A natural framework to capture the interdependency as well as the decentralized decisions is game theory [4]. The interactions between two network systems can be viewed as a noncooperative game in which each network configures itself to achieve its own objective. The Nash equilibrium solution concept can be used to

© The Author(s), under exclusive license to Springer Nature Switzerland AG 2020
J. Chen and Q. Zhu, *A Game- and Decision-Theoretic Approach to Resilient Interdependent Network Analysis and Design*, SpringerBriefs in Control, Automation and Robotics, https://doi.org/10.1007/978-3-030-23444-7_3

Fig. 3.1 Interdependent networks G including communication (G_1) and power(G_2) networks. Links in G_1 and G_2 are called inter-links, and links between G_1 and G_2 are called intra-links

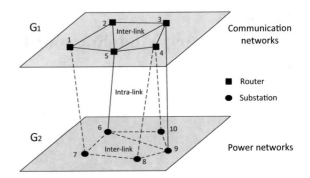

characterize and predict the outcome of such strategic interactions, which provide the analytical basis for designing mechanisms to build interdependencies that yield desirable equilibrium networks.

We establish a two-player interdependent network formation game, where two designers configure or rewire their network to achieve high connectivity of the global network. The network connectivity can be measured by *algebraic connectivity* or *Fielder value* [5], which is the second smallest eigenvalue of the Laplacian matrix of a graph. As shown in Fig. 3.1, links in the global network are classified into two types: inter-links which connect nodes in individual networks, and intra-links that connect nodes between networks. Each designer can add or remove inter-links within his network, and intra-links between the two networks to optimize the algebraic connectivity subject to the budget constraints. Despite the fact that the goals of the players are aligned, the lack of communications and coordination leads to a noncooperative game in which individual players make decisions subject to their constraints. Our network formation game problem involves two individual optimization problems which can be further rounded as a semidefinite program (SDP). We propose an algorithm with dynamic alternating plays. At every step, a player reconfigures its network based on the network obtained from the other players' strategy in the previous step. We show that the iterative algorithm converges to a Nash equilibrium after a finite number of steps.

As a comparison, we study the constrained team problem associated with the game, where two players cooperatively optimize the global interdependent network. We propose the *loss of connectivity* metric to quantify the inefficiency gap between the two problems, and use it as a guideline for increasing the efficiency of the Nash equilibrium network. We use the interdependency between a power system network and a communication network as a case study to illustrate the two-player game problem and corroborate our results.

The established interdependent network formation framework is general and can be applied to various scenarios including water networks, smart grids and transportation networks [2, 6–9]. In addition to improving the system resilience by focusing on the network topology optimization, the system operator can design another layer of artificial intelligence-based operational strategies for enhancing interdependent infrastructures resilience [10–14].

3.1.1 Problem Formulation

3.1.1.1 Static Interdependent Network Formation Game

Consider the following network formation problem: a network designer aims to optimize the network topology to achieve an objective of minimum cost and/or highest utility. Specifically, we investigate an interdependent network formation problem between two network designers. Each of them forms inter-links and intra-links to achieve their individual objectives. This lack of coordination can be captured by a game-theoretic model, and its outcome can be predicted by Nash equilibrium solutions. In the following, we describe the static interdependent network formation game.

Let P_1 and P_2 be two network designers or players involved in the game. Each player aims to add or remove links in G with the same goal of achieving high connectivity of the network. Specifically, P_1 can modify the link sets E_1 and E_{12}, and P_2 can modify E_2 and E_{21}. Here, E_{12} and E_{21} refer to the same set of intra-links of G. In addition, each player has complete knowledge of the game, and their objectives are aligned. However, without communication, they need to make individual decisions rather than resort to a centralized planner. This type of games often exists in interdependent infrastructures. For example, at the post-disaster recovery, the power systems and transportations make individual recovery plans. The insufficient communications lead to a slow restoration process despite the fact they have the same goal of fast recovery.

The players P_1 and P_2 update or rewire the network G through adding and removing a set of links. Let $\mathcal{A}_i \subseteq E_i \cup E_{ij}, i, j \in \{1, 2\}, i \neq j$, be the set of possible links P_i can add, and $x_i^e \in \{0, 1\}, i \in \{1, 2\}, e \in \mathcal{A}_i$, be the binary decision variable of P_i with $x_i^e = 1$ indicating that link e is added by P_i, and $x_i^e = 0$ indicating otherwise. Similarly, $\mathcal{D}_i \subseteq E_i \cup E_{ij}, i, j \in \{1, 2\}, i \neq j$, be the set of possible links P_i can remove, and $x_i^{e'} \in \{0, 1\}, i \in \{1, 2\}, e' \in \mathcal{D}_i$, be the binary decision variable of P_i with $x_i^{e'} = 1$ indicating that link $e' \in \mathcal{D}_i$ is removed by P_i, and $x_i^{e'} = 0$ indicating otherwise.

Both players have a budget constraint on the operating links. Since each link $e \in \mathcal{A}_i \cup \mathcal{D}_i$ in the network is associated with a cost $c_i^e \in \mathbb{R}_+, i \in \{1, 2\}$. For example, in the power and communication networks shown in Fig. 3.1, setting up PLC or wireless channels for information exchange induces a cost for network operators. Forming more links results in higher costs. Therefore, a budget for players is necessary to obtain nontrivial solutions. In the network formation game, the total budget for P_i is $M_i \in \mathbb{Z}_+$, i.e.,

$$\mathbf{c}_{i,a}^T \mathbf{x}_{i,a} + \mathbf{c}_{i,d}^T (\mathbf{1} - \mathbf{x}_{i,d}) \leq M_i, \text{ for } i \in \{1, 2\},$$

where $\mathbf{c}_{i,a} = [c_i^e]_{e \in \mathcal{A}_i} \in \mathbb{R}_+^{|\mathcal{A}_i|}$, $\mathbf{c}_{i,d} = [c_i^{e'}]_{e' \in \mathcal{D}_i} \in \mathbb{R}_+^{|\mathcal{D}_i|}$, $\mathbf{x}_{i,a} = [x_i^e]_{e \in \mathcal{A}_i} \in \{0, 1\}^{|\mathcal{A}_i|}$, $\mathbf{x}_{i,d} = [x_i^{e'}]_{e' \in \mathcal{D}_i} \in \{0, 1\}^{|\mathcal{D}_i|}$, and $\mathbf{1}$ is a $|\mathcal{D}_i|$-dimensional vector will all one entries. Here, $\mathbf{c}_{i,a}$ and $\mathbf{x}_{i,a}$ have the same order of indexing e with respect to the set $\mathcal{A}_i$.

Similarly, $\mathbf{c}_{i,d}$ and $\mathbf{x}_{i,d}$ have the same indexing as the set $\mathcal{D}_i$ with respect to e'. Another assumption is that each player updates the network topology through rewiring, then the number of added links should be equal to the number of removed links, and the total number of links stays the same. This fact can be captured by imposing the constraint

$$\sum_{e \in \mathcal{A}_i} x_i^e = \sum_{e' \in \mathcal{D}_i} x_i^{e'} .$$

This assumption is reasonable for the purpose of finding the equilibrium topology of interdependent networks. Comparing with the adding strategy in network formation [15], the rewiring method gives more insights on the resources allocation planning. Note that in this interdependent network formation game, $\mathcal{A}_i \cap \mathcal{D}_i = \emptyset$, and $\mathcal{A}_i \cup \mathcal{D}_i = E_i \cup E_{ij}$ for $i \in \{1, 2\}$, since we do not allow both players to add or remove the same. This condition can be naturally satisfied by the iterative algorithm described in later in Sect. 3.1.3.

The decision variables are $\mathbf{x}_i = [\mathbf{x}_{i,a}, \mathbf{x}_{i,d}]$ for player P_i. Denote by F_1 and F_2 the action spaces of P_1 and P_2, respectively, which are sets of all the possible link formation strategies for the players within the budget constraints. The goal of both players is to maximize the algebraic connectivity of the global network G. Hence, the utility function of both players is described by $\lambda_2(\mathbf{x}_1, \mathbf{x}_2) : F_1 \times F_2 \rightarrow \mathbb{R}_+$. More details about algebraic connectivity can be found in Sect. 2.2. We then can summarize into the following optimization problem $\mathcal{Q}_i$ for P_i for $i = 1, 2$:

$$\mathcal{Q}_i : \quad \max_{\mathbf{x}_i} \ \lambda_2 \left(\mathbf{L} + \sum_{e \in \mathcal{A}_i} x_i^e \mathbf{a}_e \mathbf{a}_e^T - \sum_{e' \in \mathcal{D}_i} x_i^{e'} \mathbf{d}_{e'} \mathbf{d}_{e'}^T \right)$$

$$\text{s.t.} \quad \sum_{e \in \mathcal{A}_i} x_i^e = \sum_{e' \in \mathcal{D}_i} x_i^{e'} ,$$

$$\mathbf{c}_{i,a}^T \mathbf{x}_{i,a} + \mathbf{c}_{i,d}^T (\mathbf{1} - \mathbf{x}_{i,d}) \le M_i ,$$

$$\mathbf{1}^T \mathbf{x}_{i,a} + \mathbf{1}^T (\mathbf{1} - \mathbf{x}_{i,d}) = h_i ,$$

$$\mathbf{x}_{i,a} = [x_i^e]_{e \in \mathcal{A}_i} \in \{0, 1\}^{|\mathcal{A}_i|} ,$$

$$\mathbf{x}_{i,d} = [x_i^{e'}]_{e' \in \mathcal{D}_i} \in \{0, 1\}^{|\mathcal{D}_i|} ,$$

where $\mathbf{a}_e$ is the added link vector whose formal definition is presented in (2.7); $\mathbf{d}_{e'}$ is the deleted link vector where $\mathbf{d}_{e'}(i) = 1$, $\mathbf{d}_{e'}(j) = -1$ and other entries zero for link e' connecting nodes i and j; $\mathbf{1}$ is a column vector with all ones of appropriate dimension; and h_i is the total number of links that player i can form.

Following Sect. 2.1, the interdependent network formation game can be represented by a triplet $\mathcal{G} := \{\mathcal{N}, \{F_i\}_{i \in \mathcal{N}}, \lambda_2\}$, where $\mathcal{N} := \{1, 2\}$ is the set of the players, F_i are the action spaces. Since both players have an aligned objective function given by λ_2, game $\mathcal{G}$ is characterized to a potential game. A natural solution concept to this game is Nash equilibrium, which has been introduced in Chap. 2.

Definition 3.1 A strategy profile $(\mathbf{x}_1^*, \mathbf{x}_2^*)$ is a Nash equilibrium of the game $\mathcal{G}$ if for every $\mathbf{x}_i \in F_i, i \in \mathcal{N}$,

$$\lambda_2(\mathbf{x}_1^*, \mathbf{x}_2^*) \geq \lambda_2(\mathbf{x}_1, \mathbf{x}_2^*),$$
$$\lambda_2(\mathbf{x}_1^*, \mathbf{x}_2^*) \geq \lambda_2(\mathbf{x}_1^*, \mathbf{x}_2).$$

At equilibrium, no player can increase the network connectivity individually by changing his rewiring strategy.

3.1.1.2 Jointly Constrained Team Problem

With sufficient communications and coordinations, the network formation game problem leads to a *constrained team problem* in which P_1 and P_2 jointly optimize the algebraic connectivity. In this problem, two players share the same objective while they should consider the constraints of both players together. Given an initial network and its Laplacian $\mathbf{L}$, the constrained team problem $\mathcal{TP}$ can be written as:

$$\mathcal{TP}: \quad \max_{\mathbf{x}_1, \mathbf{x}_2} \lambda_2\left(\mathbf{L} + \sum_{i=1}^{2}\sum_{e\in\mathcal{A}_i} x_i^e \mathbf{a}_e \mathbf{a}_e^T - \sum_{i=1}^{2}\sum_{e'\in\mathcal{D}_i} x_i^{e'} \mathbf{d}_{e'} \mathbf{d}_{e'}^T\right)$$

$$\text{s.t.} \quad \mathbf{c}_{i,a}^T \mathbf{x}_{i,a} + \mathbf{c}_{i,d}^T(\mathbf{1} - \mathbf{x}_{i,d}) \leq M_i,$$

$$\mathbf{1}^T \mathbf{x}_{i,a} + \mathbf{1}^T(\mathbf{1} - \mathbf{x}_{i,d}) = h_i,$$

$$\sum_{e\in\mathcal{A}_1} x_i^e = \sum_{e'\in\mathcal{D}_1} x_i^{e'},$$

$$\mathbf{x}_{i,a} \in \{0, 1\}^{|\mathcal{A}_i|}, \ \mathbf{x}_{i,d} \in \{0, 1\}^{|\mathcal{D}_i|}, \ i = 1, 2.$$

Due to the collective efforts, the solution to $\mathcal{TP}$ denoted by $(\mathbf{x}_1^o, \mathbf{x}_2^o)$ usually yields a higher connectivity than its Nash equilibrium counterpart. We measure this connectivity gap by using the *loss of connectivity* (LOC) metric which is defined as

$$\text{LOC}(\mathbf{x}_1^*, \mathbf{x}_2^*) := \frac{\lambda_2(\mathbf{x}_1^o, \mathbf{x}_2^o) - \lambda_2(\mathbf{x}_1^*, \mathbf{x}_2^*)}{\lambda_2(\mathbf{x}_1^o, \mathbf{x}_2^o)}.$$

Note that LOC is between 0 and 1, and it can inform network designers how bad the system performance is when there is lack of coordination, and provides a guideline for designers to close the gap between team and Nash equilibrium solutions.

3.1.2 Nash Equilibrium Analysis

In this section, we analyze the feasibility and existence of Nash equilibrium of the interdependent network formation game formulated in Sect. 3.1.1.

3.1.2.1 Feasibility

The first step is to understand whether there exists a feasible solution to $\mathcal{Q}_i$. We consider the following constraints on the budgets of each player.

Lemma 3.1 *Let $s(j)$ be the jth largest value in the vector $\mathbf{c}_{i,a}$. Then, $\mathcal{Q}_i$ is feasible if $\sum_{j=1}^{h_i} c_i^{s(j)} \leq M_i$, for $i = 1, 2$.*

3.1.2.2 Existence of Nash Equilibrium

Note that each player's action set is finite for given (n_1, n_2), (h_1, h_2) and (M_1, M_2). For all possible action profiles, we care about the existence of Nash equilibrium of this game.

Theorem 3.1 *There exists at least one pure Nash equilibrium in the interdependent network formation game.*

Proof This network formation game can be written in a matrix form, and both players share the same objective. Thus, there exists at least one $(\mathbf{x}_1^*, \mathbf{x}_2^*)$ that jointly optimizes the network connectivity. Then, $\lambda_2(\mathbf{x}_1^*, \mathbf{x}_2^*) \geq \lambda_2(\mathbf{x}_1, \mathbf{x}_2^*)$, $\forall \mathbf{x}_1 \in F_1$, $\lambda_2(\mathbf{x}_1^*, \mathbf{x}_2^*) \geq \lambda_2(\mathbf{x}_1^*, \mathbf{x}_2)$, $\forall \mathbf{x}_2 \in F_2$, where F_1 and F_2 are action spaces of P_1 and P_2, respectively. Since $(\mathbf{x}_1^*, \mathbf{x}_2^*)$ is a dominant strategy of the players, then $(\mathbf{x}_1^*, \mathbf{x}_2^*)$ is a pure Nash equilibrium in this game. $\qquad\square$

Remark 3.1 The existence of a mixed-strategy Nash equilibrium is ensured by Nash's Theorem 2.1. The equilibria of interest here are pure ones.

Corollary 3.1 *The jointly constrained team optimal solution $(\mathbf{x}_1^o, \mathbf{x}_2^o)$ also constitutes a Nash equilibrium of the network formation game $\mathcal{G}$.*

Proof First, the solution to $\mathcal{TP}$ satisfies $(\mathbf{x}_1^o, \mathbf{x}_2^o) \in F_1 \times F_2$ because of the joint constraints in $\mathcal{TP}$. If $(\mathbf{x}_1^o, \mathbf{x}_2^o)$ is not a Nash equilibrium, then, there exists $\mathbf{x}_1 \in F_1$, such that $\lambda_2(\mathbf{x}_1^o, \mathbf{x}_2^o) < \lambda_2(\mathbf{x}_1, \mathbf{x}_2^o)$, or there exists $\mathbf{x}_2 \in F_2$, such that $\lambda_2(\mathbf{x}_1^o, \mathbf{x}_2^o) < \lambda_2(\mathbf{x}_1^o, \mathbf{x}_2)$. In this sense, $(\mathbf{x}_1^o, \mathbf{x}_2^o)$ is not an optimal solution to $\mathcal{TP}$ which contradicts the assumption. Hence, $(\mathbf{x}_1^o, \mathbf{x}_2^o)$ should be a Nash equilibrium solution. $\qquad\square$

3.1.3 Algorithm Design

In this section, we present a best-response alternating play algorithm to obtain the Nash equilibrium solution, and study its convergence property.

3.1.3.1 Best-Response Alternating Play Mechanism

In practice, the Nash equilibrium can be interpreted as an outcome of iterative algorithms. Consider two players who play the game $\mathcal{G}$ by taking turns to update their strategies. Given an initial configuration of the network G_{k-1} at time $k-1$, at time step $k \in \mathbb{Z}_+$, P_1 solves the optimization $\mathcal{Q}_{1,k}$ by choosing $\mathbf{x}_{1,k} = [\mathbf{x}_{1,a}^k, \mathbf{x}_{1,d}^k]$. Due to this rewiring of links of P_1, the network topology changes from G_{k-1} to G_k, with the inter-links of G_1 and the intra-links modified. However, the inter-links of G_2 remain intact. Note that we have introduced index k to the notations to indicate specifically the problem to be solved at time k. For example, $\mathcal{Q}_{1,k}$ is the problem that P_1 solves at time k, and $\mathbf{x}_{1,k}$ is the corresponding solution. At time $k+1$, P_2 finds the optimal solution $\mathbf{x}_{2,k+1} = [\mathbf{x}_{2,a}^{k+1}, \mathbf{x}_{2,d}^{k+1}]$ to the problem $\mathcal{Q}_{2,k+1}$ based on the topology G_k. This iteration leads to the change of the inter-links of G_2 and the intra-links, but the inter-links of G_1 remain intact. This process can be iteratively proceed until no rewiring is possible, and the solution obtained in the end corresponds to the Nash equilibrium of the interdependent network formation game in which no players can deviate from their strategy unilaterally to achieve a better payoff.

It is evident that this iterative scheme automatically satisfies the condition that both players cannot add or remove the same link at the same time during each update. Moreover, this algorithm is often observed in practice. The transit authority of New York City will respond to the power system network failure to reroute its traffic, while the power network will reroute the power supply in response to the emergency. This Nash equilibrium can be used to predict the outcome of two network operators when they have limited and insufficient communications when disasters hit the city.

3.1.3.2 Convergence of the Algorithm

During the network formation game, one player's optimal strategy at each round of the update is the best response to the strategy of the other player from the previous round. For convenience, we define $-i := \{1, 2\}\backslash\{i\}$, where $i \in \{1, 2\}$, to denote the player other than i. Then, we have the following theorem about the outcome of updates.

Theorem 3.2 *Under the best-response alternating play algorithm, the algebraic connectivity of G converges to a pure Nash equilibrium after a finite number of steps.*

Proof Since P_1 and P_2 each can form a finite number of links in the game, the optimal algebraic connectivity of G is upper bounded. Denote players' action at step k as $\mathbf{x}_{1,k}$ and $\mathbf{x}_{2,k}$, respectively. Then, $\lambda_2(\mathbf{x}_{1,k+1}, \mathbf{x}_{2,k+1}) \geq \lambda_2(\mathbf{x}_{1,k}, \mathbf{x}_{2,k})$, for all $k \in \mathbb{Z}_+$, which indicates that the outcome of each step is a non-decreasing sequence. Since the action spaces of players are finite, then, players cannot increase the λ_2 of G continually by reforming the network. Therefore, $\exists k \in \mathbb{Z}_+$, such that the strategy profile $(\mathbf{x}_{1,k+1}, \mathbf{x}_{2,k+1})$ and $(\mathbf{x}_{1,k}, \mathbf{x}_{2,k})$ are the same, and $\lambda_2(\mathbf{x}_{1,k+1}, \mathbf{x}_{2,k+1}) = \lambda_2(\mathbf{x}_{1,k}, \mathbf{x}_{2,k})$. Since P_i's strategy at step $k+1$ is the best response to the network that formed after

P_{-i}'s action at step k, then, $\lambda_2(\mathbf{x}_{1,k'}, \mathbf{x}_{2,k'}) = \lambda_2(\mathbf{x}_{1,k'-1}, \mathbf{x}_{2,k'-1})$, $\forall\, k' > k$. Since the utility function is non-decreasing during the game, then, we know that the action profile $(\mathbf{x}_{1,k'}, \mathbf{x}_{2,k'})$ will not change, and the λ_2 stays constant for $k' > k$. Hence, the network reaches an equilibrium after a finite number of k steps, and the strategy profile $(\mathbf{x}_{1,k}, \mathbf{x}_{2,k})$ is a Nash equilibrium. We denote the sequence of λ_2 in this game as $\{v(k)\}$, and its upper bound as $\bar{v}$. Then, $\forall \epsilon > 0$, $|v(k') - \bar{v}| = 0 < \epsilon$, $\forall k' \geq k$, which results in the convergence of the algebraic connectivity in the game.   □

3.1.4 SDP-Based Approach

In this section, we propose a semidefinite programming approach to solve the optimization problem $\mathcal{Q}_i$.

3.1.5 Alternative Problem Formulation

To simplify the formulation of the original optimization problem $\mathcal{Q}_i$, we first introduce the following lemma.

Lemma 3.2 *At step k, P_i will not remove those intra-links formed by P_{-i} at the previous step $k - 1$.*

Proof Without loss of generality, we assume that each link has the same cost, and P_1 is playing the game. Suppose P_1 removes a set of intra-links B formed by P_2 at step $k - 1$. Denote the optimal network formed after P_1's action as G_k, and the current strategy profile of players as $(\mathbf{x}_{1,k}, \mathbf{x}_{2,k-1})$. If P_1 does not remove links in set B, then he can form $|B|$ additional links based on G_k, and we denote the strategy of players at this case as $(\mathbf{x}'_{1,k}, \mathbf{x}_{2,k-1})$. Since algebraic connectivity is non-decreasing on the link addition with the same set of nodes [5], then we obtain $\lambda_2(\mathbf{x}'_{1,k}, \mathbf{x}_{2,k-1}) \geq \lambda_2(\mathbf{x}_{1,k}, \mathbf{x}_{2,k-1})$. Therefore, P_i will not remove intra-links that formed by P_{-i} at the previous step.   □

Based on Lemma 3.2, we can reformulate $\mathcal{Q}_i$ as follows.

Corollary 3.2 *Links rewiring problem $\mathcal{Q}_i$ is equivalent to the following links adding problem for $i = 1, 2$:*

$$\mathcal{Q}'_i: \quad \max_{\mathbf{x}_i} \quad \lambda_2\left(\mathbf{L}' + \sum_{e \in \mathcal{T}_i} x_i^e \mathbf{l}_e \mathbf{l}_e^T\right)$$
$$\text{s.t.} \quad \mathbf{c}_i^T \mathbf{x}_i \leq M_i,$$
$$\mathbf{1}^T \mathbf{x}_i = h_i,$$
$$\mathbf{x}_i \in \{0, 1\}^{|\mathcal{T}_i|},$$

where $\mathbf{L}'$ *is the Laplacian matrix of G which is only contributed by the previous step of* P_{-i}. $\mathcal{T}_i$ *is the link set that contains all inter-links in* G_i, *and all intra-links except those already formed by* P_{-i}, *i.e.,* $\mathcal{T}_i = \mathcal{A}_i \cup \mathcal{D}_i \backslash \mathcal{D}_{-i}$. $\mathbf{x}_i$, $\mathbf{l}_e$ *and* $\mathbf{c}_i$ *have the same interpretations as* $\mathbf{x}_i$, $\mathbf{a}_e$ *and* $\mathbf{c}_i$ *in* $\mathcal{Q}_i$, *respectively, with only a slight variation on the vector size.*

Proof The interpretation of $\mathcal{Q}'_i$ is as follows: P_i removes all links that he has formed at the previous step first, then starts from scratch and forms another h_i links. Therefore, to show $\mathcal{Q}_i$ and $\mathcal{Q}'_i$ are equivalent, it is enough to show that P_i will not remove intra-links that formed by P_{-i} at the previous step. This is true which is proved in Lemma 3.2. □

3.1.5.1 Nonzero-Sum Game and SDP Formulation

We define a new game $\tilde{\mathcal{G}} := \{\mathcal{N}, \{\tilde{F}_i\}_{i \in \mathcal{N}}, \{\alpha_i\}_{i \in \mathcal{N}}\}$, where $\mathcal{N} = \{1, 2\}$ is the set of the players, $\tilde{F}_i$ are the feasible action spaces defined by problem $\mathcal{Q}'_i$ in Corollary 3.2, and α_i is the objective for player i. Then, we have the following.

Proposition 3.1 *The nonzero-sum game* $\tilde{\mathcal{G}}$ *is equivalent to the potential game* $\mathcal{G}$ *defined in Sect. 3.1.1.1, and its formulation can be captured by the following SDP problem:*

$$\mathcal{Q}''_i : \max_{\mathbf{x}_i, \alpha_i} \quad \alpha_i$$

$$\text{s.t.} \quad \mathbf{L}' + \sum_{e \in \mathcal{T}_i} x_i^e \mathbf{l}_e \mathbf{l}_e^T \succeq \alpha_i \left(\mathbf{I}_n - \frac{1}{n} \mathbf{1} \mathbf{1}^T \right),$$

$$\mathbf{c}_i^T \mathbf{x}_i \leq M_i,$$

$$\mathbf{1}^T \mathbf{x}_i = h_i,$$

$$\mathbf{x}_i \in \{0, 1\}^{|\mathcal{T}_i|}.$$

where $\alpha_i \in \mathbb{R}$, *for* $i = 1, 2$, *and* $\mathbf{I}_n$ *is an n-dimensional identity matrix.*

Proof We have shown that $\mathcal{Q}'_i$ is equivalent to $\mathcal{Q}_i$ in Corollary 3.2. Then, to show game $\tilde{\mathcal{G}}$ is equivalent to game $\mathcal{G}$, it is sufficient to show that its SDP formulation is equivalent to the formulation of $\mathcal{Q}'_i$. Based on the result in [16], i.e., the optimization problem $\max_{\mathbf{x}} \lambda_2(\mathbf{L}(\mathbf{x}))$ is equivalent to $\max_{\mathbf{x}, \beta} \beta$ while subjecting to $\mathbf{L}(\mathbf{x}) \succeq \beta(\mathbf{I}_n - \frac{1}{n}\mathbf{1}\mathbf{1}^T)$, where β is a scalar. Therefore, the equivalence of $\mathcal{G}$ and $\tilde{\mathcal{G}}$ follows from the fact that $\mathcal{Q}'_i$ and $\mathcal{Q}''_i$ are equivalent. □

Note that game $\tilde{\mathcal{G}}$ enables an efficient computation of best-response algorithms. It, however, is a nonzero-sum game since the objective functions in $\mathcal{Q}''_i$ are different for players $i = 1, 2$. Based on Proposition 3.1, we know that $\tilde{\mathcal{G}}$ admits at least one pure Nash equilibrium, and its best-response algorithm is similar to the one in Sect. 3.1.3.

Since the elements in the decision vector x_i in $\mathcal{Q}_i''$ is binary, we denote this problem as a binary game problem (BGP). BGP is NP-hard and not easy to solve. By changing the boolean constraint $\mathbf{x}_i \in \{0, 1\}^{|\mathcal{T}_i|}$ to the linear constraint $\mathbf{x}_i \in [0, 1]^{|\mathcal{T}_i|}$, we obtain a relaxed game problem (RGP) as

$$\mathcal{Q}_i^R : \max_{\mathbf{x}_i, \alpha_i} \quad \alpha_i$$

$$\text{s.t.} \quad \mathbf{L}' + \sum_{e \in \mathcal{T}_i} x_i^e \mathbf{l}_e \mathbf{l}_e^T \succeq \alpha_i \left(\mathbf{I}_n - \frac{1}{n} \mathbf{1} \mathbf{1}^T \right),$$

$$\mathbf{c}_i^T \mathbf{x}_i \leq M_i,$$

$$\mathbf{1}^T \mathbf{x}_i = h_i,$$

$$\mathbf{x}_i \in [0, 1]^{|\mathcal{T}_i|}.$$

We have transformed $\mathcal{Q}_i$ which is hard to deal with to an easier RGP $\mathcal{Q}_i^R$, and SeDuMi [17] is adopted to solve the RGP. Moreover, the solution to the RGP $\mathcal{Q}_i^R$ severs as an upper bound of the solution to BGP $\mathcal{Q}_i''$ due to the relaxation.

3.1.5.2 Rounding Techniques for RGP

Denote the strategy of players as $\mathbf{x}_i^*$ and $\mathbf{x}_{-i}^*$, respectively. Then, several rounding methods are listed below.

- *Greedy*: Selecting the h_i and h_{-i} largest entries in $\mathbf{x}_i^*$ and $\mathbf{x}_{-i}^*$ alternatively at one shot.
- *Link-by-Link*: We select the largest entry that is smaller than 1 in $\mathbf{x}_i^*$, and set it to 1. Then, update the RGP by incorporating this link. Next, solve the new RGP again, and repeat these two steps for h_i times. Same for $\mathbf{x}_{-i}^*$.
- *Log Link-by-Link*: The only difference with the link-by-link method is that we choose the best half of the remaining elements rather than one at a time. Then, only around $\log(h_i)$ and $\log(h_{-i})$ number of RGPs need to be solved for P_i and P_{-i} at every step, respectively.

3.1.6 Case Studies

3.1.6.1 Interdependent Networks Model

We consider the communication network A and power network B that contain 3 routers and 5 substations, respectively. Both network managers are able to set up wireless links between routers and substations known as intra-links. Inside the network, A forms wireless inter-links, while B sets up inter-links via PLC. Note that the switching of PLC over a link inside the power network is enabled by the flexible AC transmission system (FACTS) technology [18]. In addition, the cost of each link

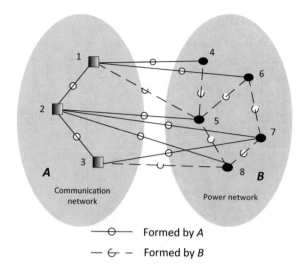

Fig. 3.2 Interdependent power and communication networks. 8 links are contributed by network A, and 7 links are formed by network B

is related to its type, i.e., inter-link or intra-link, and PLC or wireless communication. For simplicity, we normalize the budget and the cost of communication links. Specifically, each inter-link costs 1 and 1.2 units in networks A and B, respectively, and each wireless intra-link takes 1.5 units between A and B. Figure 3.2 shows the initial configuration of the interdependent networks.

3.1.6.2 Performance of the Two Formation Models

In this case study, $h_1 = 8$, $h_2 = 7$, $M_1 = 12$ and $M_2 = 10$. Figure 3.3a shows the performance of the game model and the team model on the algebraic connectivity. The network connectivity is increasing step by step using the game method, and reaches an equilibrium after 9 steps, illustrating the effectiveness of the algorithm. The final configuration of the interdependent network is shown in Fig. 3.3c. Moreover, the connectivity of the Nash equilibrium network is the same as that of the constrained team network, and thus LOC is 0.

3.1.6.3 Comparison of Different Initial Configurations

By rewiring links $(1, 2)$ to $(1, 3)$ in the initial configuration in Fig. 3.2, the best response dynamic algorithm can converge to a Nash equilibrium in 5 steps shown in Fig. 3.3b, d is the equilibrium network. Though, in this case, the final network has the same algebraic connectivity as that in Fig. 3.3a, and it is also optimal (i.e., LOC = 0), the equilibrium network topology is different for these two scenarios. Therefore, there exist more than one Nash equilibria in this interdependent network formation game, and it corroborates that the jointly team optimal solution also constitutes a Nash equilibrium.

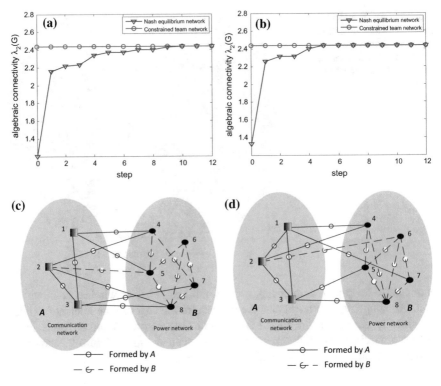

Fig. 3.3 Both **a** and **b** show the performances of the game model and team model on $\lambda_2(G)$. The initial network for **a** is shown in Fig. 3.2, and **b**'s initial network is the same as that in Fig. 3.2 except link (1, 2) is changed to (1, 3). **c** and **d** are the final equilibrium networks for cases **a** and **b**, respectively

3.1.6.4   Impact of the Budget on Network Formation Game

In previous case studies, $M_1 = 12$ and $M_2 = 10$ indicate that both network designers have a sufficiently large budget, and they can create links wherever necessary. Next, we investigate the impact of the budget on the network formation game. Assuming $M_1 = 11$ and $M_2 = 8.5$, and also the initial network configuration is the same as that in Fig. 3.2 except rewiring link (5, 1) to (4, 6). Then, the results of the game and team solutions are shown in Fig. 3.4. We can see that the obtained equilibrium network is not optimal, and the LOC is equal to 4.1%. The constrained optimal solution is the same as that in Fig. 3.3. This can be directly verified by the network in Fig. 3.3c, as it also satisfies the budget constraints $M_1 = 11$ and $M_2 = 8.5$. The reason accounting for nonzero LOC in this case is that during the network updating process, players cannot implement desired rewiring strategies due to the insufficient budget at some steps. Therefore, the achieved equilibrium is a suboptimal Nash equilibrium.

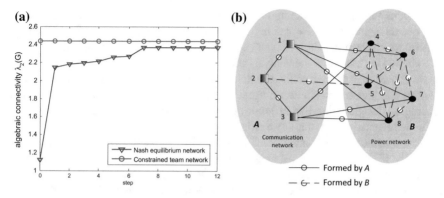

Fig. 3.4 **a** shows the performances of the game model and team model with budgets $M_1 = 11$ and $M_2 = 8.5$. **b** is the resulting equilibrium network

To sum up, with sufficient budgets, players can achieve a Nash equilibrium network same as the constrained team optimal network by using the proposed best response dynamic algorithm from any initial network configuration. When the budgets are limited, the yielding equilibrium network can be a suboptimal Nash equilibrium network comparing with its constrained team network counterparts. Therefore, cooperation is more preferable for budget scarcity cases in which the LOC can be dropped to zero.

3.2 Dynamic Network Resilience Game

The previous section has studied the connectivity of static network-of-networks. We extend the framework to a dynamic one and investigate dynamic network resilience game in this section.

Cooperative mobile autonomous system (MAS) has a wide range of applications, such as rescue and monitoring the crowd in mission critical scenarios. One of the challenges in designing the MAS network is to maintain the connectivity between agents/robots,[1] since a higher connectivity enables faster information spreading and hence a high-level of situational awareness. In [19], the authors have successfully tackled connectivity control of a single network of cooperative robots. Recent advances in networked systems have witnessed emerging applications involving multi-layer networks or network-of-networks [20–22]. For example, in the battlefield, unmanned aerial vehicles (UAVs) and unmanned ground vehicles (UGVs) execute tasks together, and the whole network can be seen as a two-layer interdependent network [23]. The connectivity of the two-layer network can play a key role in the

---

[1]The "agent" refers to the robot in our MAS network. We also use the terms "MAS network" and "robotic network" interchangeably.

operations for a given collaborative mission. To enable the real-time decision making of each agent, the integrated network needs to guarantee a level of connectivity. Therefore, the current single network control paradigm is not yet sufficient to address the challenges related to the analysis and design of multi-layer MAS networks. The main objective in the rest of the chapter is to develop a theoretic framework that can capture the interactions between agents within a network and across networks and enable the design of distributed control algorithms that can maintain the connectivity in both adversarial and non-adversarial environment.

The operator of each layer MAS network aims to maximize the algebraic connectivity [5] of the global network. If the whole network is fully cooperative or governed by a single network operator, then the designed network is a *team-optimal* solution. However, in practice, different layers of robotic networks are often operated by different entities, which makes the coordination between separate entities difficult. This uncoordinated control design naturally leads to a *system-of-systems* (SoS) framework of the multi-layer MAS network. For example, in the aforementioned two-layer UAV and UGV mobile networks, though the objectives of two network operators are aligned, the UAVs are operated by one entity while the UGVs is operated by another. The lack of the centralized planning can result in insufficient coordination between two networks and lead to disruptions in connectivity and security vulnerabilities. To address this problem, we establish a *Nash* game-theoretic model in which two players, i.e., network operators, control robots at their layer, to maximize the global connectivity independently. This model captures the lack of coordination between players and their decentralized decision making in optimizing the SoS performance.

Cybersecurity is another critical concern of networked systems because of the integration with IoT components [24, 25]. To address the security issues in multilayer systems, a holistic approach capturing the interdependencies is necessary [26, 27]. In the MAS, the agents are prone to adversarial attacks, e.g., the communication links between robots can be jammed (e.g., [28]) which decreases the connectivity. Therefore, secure control of the multi-layer MAS networks is critical to maintain the SoS performance at a high level. To this end, we model the mobility of robots by taking into account the imperfect communication links under adversarial environment. Specifically, each network operator anticipates the jamming attacks and controls robots by anticipating that a set of critical links between agents can be compromised. This secure control design can be modeled by a *Stackelberg* game between each network operator and the attacker.

We integrate the modeled Nash game between two network operators as well as the Stackelberg game between the network operator and the attacker which further yields a *games-in-games* framework. This new type of game provides a holistic modeling that integrates the network-network interactions and the agent-adversary interactions together for the secure and decentralized control design of multi-layer MAS networks. We propose a *meta-equilibrium* solution for this games-in-games which includes the optimal strategy of network operators and the strategic jamming attacks of adversary. We further develop a resilient and decentralized mechanism that guides the online design of MAS for achieving a meta-equilibrium solution. A typical example is that two network operators update their own network in a round-robin fashion based on the current network to maximize the network connectivity. This alternating-play mechanism induces an iterative algorithm that can converge to a meta-equilibrium asymptotically.

The games-in-games framework provides a theoretical foundation for understanding the agile resilience of the system to cyberattacks, which is a critical system property for the MAS network to recover quickly especially for mission-critical applications. When robots or communication links in the network are compromised, the integrated MAS network under the designed control strategy needs to respond to the unexpected disruptions with agility to mitigate the loss of connectivity. Hence, to investigate the resilience of the designed algorithm, we first introduce the GPS spoofing attacks. The resilience is measured by the enhanced SoS performance through the design of post-attack control strategies. Simulation results show that the multi-layer MAS network is resilient to attacks using the proposed control method. After the detection of GPS spoofing attack by the network operator, the MAS network shows agile resilience to the attack and the system can adapt and reconfigure itself to an efficient meta-equilibrium that coincides with the one without attack.

Related Work

MAS has been applied to a number of emerging fields. One of them is drone delivery [29]. Another example is the unmanned aerial vehicles (UAVs) assisted sensing and communication networks for disaster response and recovery [30]. Since we focus on controlling groups of MAS, this work is also related to the classical control of multi-agent systems [31–33].

One critical factor needs to be considered for MAS network is its connectivity. When MAS is adopted in mission-critical scenarios, such as battlefields and disaster-affected areas, a higher network connectivity provides a higher level of situational awareness [34]. Maximizing the algebraic connectivity of networks has been investigated extensively in literature, including single-layer static network [15], single-layer mobile network [31, 35], multi-layer static network [21, 36, 37]. We focus on optimizing a new category of multi-layer MAS network connectivity in this section.

3.2.1 Games-in-Games Framework

We introduce the system framework which includes the wireless communication model and the strategic interdependent MAS network formation.

3.2.1.1 Wireless Communication Model

In the MAS, we consider a set V of robots in the network, and their positions at time k are defined by the vector $\mathbf{x}(k) = \big(x_1(k);\ x_2(k);\ \dots;\ x_n(k)\big) \in \mathbb{R}^{3n}$. Robots in the same network can exchange data via wireless communications. Denote the communication link between robots i and j as (i, j). Then, the strength of the communication

Fig. 3.5 Communication strength under function $f(d) = \delta^{(c_1-d)/(c_1-c_2)}$ with $\delta = 0.1, c_1 = 2$ and $c_2 = 6$

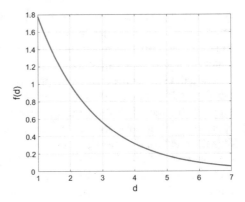

link (i, j) is similar to the weight of the link in a network. Thus, we associate a weight function $w : \mathbb{R}^3 \times \mathbb{R}^3 \to \mathbb{R}_+$ with every communication link (i, j), such that

$$w_{ij}(k) = w\big(x_i(k), x_j(k)\big) = f\left(||x_{ij}(k)||_2^2\right), \tag{3.1}$$

for some differentiable $f : \mathbb{R}_+ \to \mathbb{R}_+$, where $x_{ij}(k) := x_i(k) - x_j(k)$, and $||x_{ij}(k)||_2$ is the distance between robots i and j. To capture the communication strength decay with the distance, f is a monotonically decreasing function. A typical choice of f is $f(d) = \delta^{(c_1-d)/(c_1-c_2)}$, where δ, c_1 and c_2 are positive constants. Note that different forms of f capture various decay rates of communication strength with distance [38]. Thus, the weight of the link between robots is positive if their distance is within a threshold and degenerates to zero otherwise. Figure 3.5 shows an example of f with $\delta = 0.1, c_1 = 2$ and $c_2 = 6$.

3.2.1.2   Secure Interdependent MAS Network Formation

A two-layer MAS network model is shown in Fig. 3.6, where networks G_1 and G_2 include n_1 and n_2 number of robots, respectively. More generally, we label robots in G_1 as $1, 2, \ldots, n_1$, and robots in G_2 as $n_1 + 1, n_1 + 2, \ldots, n_1 + n_2$, i.e., $V_1 := \{1, 2, \ldots, n_1\}$ and $V_2 := \{n_1 + 1, n_1 + 2, \ldots, n\}$. Note that $n = n_1 + n_2$. Robots in these two layers can also communicate, and this kind of communication link is called *intra-link* while the link inside of a network is known as *inter-link*. Exchanging data between robots at different layers can be more difficult than that of the robots at the same layer because of possible larger distance. In this situation, to enable information exchange between networks, we can assume that intra-link has a smaller communication strength decay comparing with that of inter-link by choosing different f in (3.1). For notational clarity, we adopt the same communication strength function for intra-links and inter-links. The agents at two layers are interdependent, and thus the integrated MAS network can be modeled as a *system-of-systems*. A more detailed modeling of system-of-systems can refer to Sect. 2.2.2.

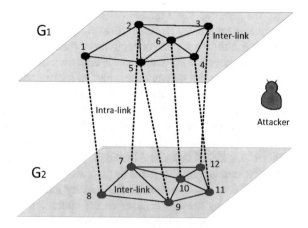

Fig. 3.6 Multi-layer MAS network in an adversarial environment

Network Designer

We consider two players, player 1 (P_1) and player 2 (P_2), operating two inter-dependent MAS networks. P_1 controls robots in network G_1, and P_2 controls robots in G_2. Specifically, P_1 and P_2 update their own network with a fixed frequency by controlling the positions of robots. After each update, the communication link strength between robots are modified due to the change of distance. For simplicity, define $-\gamma := \{1, 2\}\backslash\gamma$, where $\gamma \in \{1, 2\}$, and $\mathbf{x} := (\mathbf{x}_1, \mathbf{x}_2)$, where $\mathbf{x}_1 := (x_1; \ldots; x_{n_1}) \in \mathbb{R}^{3n_1}$ and $\mathbf{x}_2 := (x_{n_1+1}; \ldots; x_n) \in \mathbb{R}^{3n_2}$. Specifically, $\mathbf{x}_1$ and $\mathbf{x}_2$ are decision variables denoting the position of robots in G_1 and G_2, respectively. In addition, the action spaces of P_1 and P_2 are denoted by $\mathcal{X}_1$ and $\mathcal{X}_2$, respectively, which include all the possible network configurations. The set of pure strategy profiles $\mathcal{X} := \mathcal{X}_1 \times \mathcal{X}_2$ is the Cartesian product of the individual pure strategy sets. For each update, P_γ's strategy $\mathcal{X}_\gamma$ is based on the current configuration of network $G_{-\gamma}$. The goal of both players is to optimize the SoS performance, i.e., maximize the algebraic connectivity of the global network G. Hence, the utility function for both players is $\lambda_2(\mathbf{L}_G(\mathbf{x}))\colon \mathcal{X} \to \mathbb{R}_+$, where $\mathbf{L}_G(\mathbf{x})$ is the Laplacian matrix of network G when mobile robots have position $\mathbf{x}$.

Remark 3.2 In general, the objectives of two players can be different rather than maximizing the algebraic connectivity of the global MAS network. However, in our problem setting, the two teams of robots execute tasks collaboratively, and thus they both aim to optimize the global network connectivity to improve the SoS performance.

In the adversarial network formation game, one of the constraints is the minimum distance between robots in each layer. Without this constraint, all robots at the same layer will converge to one point finally which is not a reasonable solution. Thus, we assign a minimum distance ρ_1 and ρ_2 for robots in G_1 and G_2, respectively.

Cyber Attacker

In addition to the network players P_1 and P_2, our framework also includes a malicious jamming attacker as shown in Fig. 3.6. The attacker is able to disrupt communication links via injecting large amount of spam into the channel which leads to the link breakdown eventually because of overload of the link. The attacker's objective is to minimize the algebraic connectivity of the network through compromising links. Generally, the behavior of attacker is unknown to the network operators. Therefore, it is difficult for the network designers to make optimal strategies that can achieve the best performances of the network. However, by knowing that attackers are strategic and are more prone to disrupt the critical communication links in the network, the network operators can design a secure MAS network resistant to cyberattacks. Specifically, network designers first anticipate that the attacker can compromise a number $\psi \in \mathbb{Z}^+$ of links, and then design the MAS network by taking into account the worst-case attack that leads to the most decrease of the network algebraic connectivity. Note that ψ quantifies the security level of the designed network.

Denote $\mathcal{A}$ by the action space of the attacker which is the set including all the possible single communication link removal in the network. For convenience, we denote $\mathbf{L}_G^e(\mathbf{x})$ by the Laplacian matrix of the network after removing a set of links $e \subseteq \mathcal{A}$, i.e., the network after attack is $G(V, E_1 \cup E_2 \cup E_{12}\backslash e)$, and the cardinality of e is $|e| = \psi$ quantifying the ability of attacker, where $\psi \in \mathbb{Z}^+$ is a positive integer. Denote the feasible set of e by $\mathcal{E}$. Then, the cost function of the attacker can be captured by $\Lambda(\mathbf{x}_1, \mathbf{x}_2, e) \triangleq \lambda_2(\mathbf{L}_G^e(\mathbf{x}))$, for $\Lambda : \mathcal{X}_1 \times \mathcal{X}_2 \times \mathcal{E} \to \mathbb{R}_+$.

3.2.1.3   Games-in-Games Formulation

During the MAS network formation, the interactions between two networks G_1 and G_2 can be modeled as a Nash game where both players aim to increase the global network connectivity. In addition, each network operator plays a Stackelberg game with the malicious jamming attacker. Therefore, the multi-layer MAS network formation in the adversarial environment can be characterized by a games-in-games framework [39–41] which is shown in Fig. 3.7. Since network operators are aware of the jamming attacks, the proposed model incorporates security considerations in the MAS network design. In the following, we specifically formulate the attacker's and network operators' problems, respectively.

Stackelberg Game

In the Stackelberg game, network designer is the leader, and jamming attacker is the follower. Background of Stackelberg game can be found in Sect. 2.1.3. The objective of the attacker is to minimize the algebraic connectivity of network G. We can summarize the strategic behavior of the attacker into the following problem:

$$Q_A^k : \quad \min_{e \subseteq \mathcal{A}, |e|=\psi} \lambda_2\big(\mathbf{L}_G^e(\mathbf{x}(k+1))\big). \tag{3.2}$$

On the leader side, network operator P_γ maximizes the algebraic connectivity of the network, where $\gamma \in \{1, 2\}$, and his decision can be obtained via solving the optimization problem:

Fig. 3.7 Games-in-Games framework which includes two network operators and one attacker. Both network operators prepare for the cyberattack which form a Stackelberg game with the attacker. In addition, two network operators are uncoordinated and aim to maximize the global network connectivity which create a Nash game

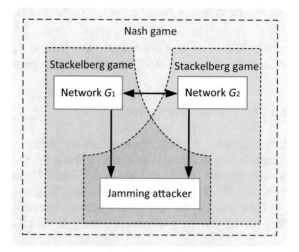

$$Q_\gamma^k : \max_{\mathbf{x}_\gamma(k+c_\gamma)} \min_{e \subseteq \mathcal{A}, |e| = \psi} \lambda_2\big(\mathbf{L}_G^e(\mathbf{x}(k+c_\gamma))\big)$$

$$\text{s.t. } ||x_{ij}(k+c_\gamma)||_2 \geq \rho_\gamma, \quad \forall (i,j) \in E_\gamma,$$
$$||x_{ij}(k+c_\gamma)||_2 \geq \rho_{12}, \quad \forall i \in V_\gamma, \ \forall j \in V_{-\gamma}, \qquad (3.3)$$
$$||x_i(k+c_\gamma) - x_i(k)||_2 \leq d_\gamma, \quad \forall i \in V_\gamma,$$
$$x_j(k+c_\gamma) = x_j(k), \ \forall j \in V_{-\gamma},$$

where $c_\gamma \in \mathbb{Z}^+$ is a positive integer indicating the update frequency; $\rho_\gamma \in \mathbb{R}_+$ is the safety distance between robots; $\rho_{12} \in \mathbb{R}_+$ is the minimum distance between robots in different layers; and $d_\gamma \in \mathbb{R}_+$ is the maximum distance that robots in network G_γ can move at each update. The constraint $x_j(k+c_\gamma) = x_j(k)$, $j \in V_{-\gamma}$ captures the uncoordinated nature that each network operator can only control the robots at his layer. Furthermore, this constraint preserves security consideration between agents in $V_{-\gamma}$ and also ensures consistent connectivity improvement when player γ updates his network.

The Stackelberg game between the attacker and network operator γ can be represented by $\Xi_\gamma := \{\mathcal{N}_\gamma, \mathcal{X}_\gamma, \mathcal{A}, \lambda_2\}$ for $\gamma \in \{1, 2\}$, where $\mathcal{N}_\gamma := \{P_\gamma, Attacker\}$ is the set of players, $\mathcal{X}_\gamma$ and $\mathcal{A}$ are action spaces and λ_2 is the objective function.

Nash Game

The interaction between two robotic networks in an adversarial environment can be characterized as a Nash game in which both players aim to increase the global network connectivity. We denote this strategic game by $\Xi_I := \{P_1, P_2, \mathcal{X}_1, \mathcal{X}_2, \lambda_2\}$.

Note that the MAS network formation game is played repeatedly over time, and its structure is the same only with different initial conditions in terms of the robots' position. This two-person interdependent MAS network formation game can be naturally generalized into an N-person game where each player controls a subset of robots in the multi-layer networks.

3.2.2 Problem Analysis and Meta-Equilibrium

In this section, we first reformulated problems in Sect. 3.2.1, and then present the solution concept of the MAS network formation game.

3.2.2.1   Problem Reformulation

Note that each network designer updates the robotic network iteratively based on the current configuration. It is essential to obtain the relationship between the updated position and the current one due to the natural dynamics of robots. To achieve this goal, we define $Z_{ij}(k) := ||x_{ij}(k)||_2^2$ for notational convenience. Analogous to applying Euler's first order method to continuous dynamics, we can obtain $Z_{ij}(k+m)$ based on the current positions $x_i(k)$ and $x_j(k)$ as follows:

$$
\begin{aligned}
Z_{ij}(k+m) + Z_{ij}(k) &= ||x_{ij}(k+m)||_2^2 + ||x_{ij}(k)||_2^2 \\
&= 2\{x_i(k+m) - x_j(k+m)\}^T \{x_i(k) - x_j(k)\}.
\end{aligned}
\tag{3.4}
$$

Similarly, by using the function f in (3.1), the updated weight $w_{ij}(k+m)$ can be expressed as:

$$
w_{ij}(k+m) = w_{ij}(k) + \left.\frac{\partial f}{\partial ||x_{ij}||_2^2}\right|_k (Z_{ij}(k+m) - Z_{ij}(k)).
\tag{3.5}
$$

Therefore, we can obtain the Laplacian matrix $\mathbf{L}_G\big(\mathbf{x}(k+m)\big)$ by using (3.5) for the global network. Specifically, its entries $\mathbf{L}_{ij}^G(k+m)$ are as follows:

$$
\mathbf{L}_{ij}^G(k+m) =
\begin{cases}
-w_{ij}(k+m), & \text{if } i \neq j,\ (i,j) \in E_1 \cup E_2; \\
-\tilde{w}_{ij}(k+m), & \text{if } i \in V_1,\ j \in V_2,\ \text{or } j \in V_1,\ i \in V_2; \\
\displaystyle\sum_{s \neq i, s \in V_1} w_{is}(k+m) + \sum_{q \neq i, q \in V_2} \tilde{w}_{iq}(k+m), & \text{if } i = j \in V_1; \\
\displaystyle\sum_{s \neq i, s \in V_2} w_{is}(k+m) + \sum_{q \neq i, q \in V_1} \tilde{w}_{iq}(k+m), & \text{if } i = j \in V_2;
\end{cases}
\tag{3.6}
$$

where w_{ij}, $\forall (i,j) \in E_1 \cup E_2$, represent the weights of inter-links inside G_1 and G_2, and $\tilde{w}_{ij}$, $\forall (i,j) \in E_{12}$, denote the weights of intra-links connecting G_1 and G_2.

Each network designer needs to solve a *max min* problem which is not straightforward to deal with. We first present the following result.

Theorem 3.3 *For a network containing n nodes, the optimization problem*

$$
\max_{\mathbf{x}} \min_{e \subseteq \mathcal{A}, |e| = \psi} \lambda_2(\mathbf{L}_G^e(\mathbf{x}))
\tag{3.7}
$$

is equivalent to

$$\max_{\mathbf{x},\beta} \quad \beta$$

$$\text{s.t. } \mathbf{L}_G^e(\mathbf{x}) \succeq \beta \left(\mathbf{I}_n - \frac{1}{n}\mathbf{1}\mathbf{1}^T\right), \; \forall e \subseteq \mathcal{A}, \; |e| = \psi, \tag{3.8}$$

where β is a scalar. Note that the optimal $\mathbf{x}$ and the corresponding objective values in these two problems are equal.

Proof Let v_i be the eigenvector associated with eigenvalue λ_i of the Laplacian matrix $\mathbf{L}_G^e(\mathbf{x})$, for $\forall i \in V$. Since $\mathbf{L}_G^e(\mathbf{x})$ is real and symmetric, its eigenvectors can be chosen such that they are real, orthogonal to each other and have norm one, i.e., $v_i^T v_j = 0, \forall i \neq j \in V$ and $v_i^T v_i = 1$. Specially, we define $v_1 := \frac{1}{\sqrt{n}}$, which is actually the eigenvector corresponding to $\lambda_1 = 0$. Then, $\mathbf{L}_G^e(\mathbf{x})$ admits a spectral decomposition of the following form

$$\mathbf{L}_G^e(\mathbf{x}) = \sum_{i=1}^n \lambda_i(\mathbf{L}_G^e(\mathbf{x})) v_i v_i^T. \tag{3.9}$$

Since $\lambda_1 = 0$, Eq. (3.9) can be simplified as

$$\mathbf{L}_G^e(\mathbf{x}) = \sum_{i=2}^n \lambda_i(\mathbf{L}_G^e(\mathbf{x})) v_i v_i^T. \tag{3.10}$$

Next, we add $\lambda_2(\mathbf{L}_G^e(\mathbf{x})) v_1 v_1^T$ to both sides of (3.10) and obtain $\mathbf{L}_G^e(\mathbf{x}) + \lambda_2(\mathbf{L}_G^e(\mathbf{x})) v_1 v_1^T = \sum_{i=2}^n \lambda_i(\mathbf{L}_G^e(\mathbf{x})) v_i v_i^T + \lambda_2(\mathbf{L}_G^e(\mathbf{x})) v_1 v_1^T$. By using property (2.8), we further have

$$\mathbf{L}_G^e(\mathbf{x}) + \lambda_2(\mathbf{L}_G^e(\mathbf{x})) v_1 v_1^T \succeq \lambda_2(\mathbf{L}_G^e(\mathbf{x})) \sum_{i=1}^n v_i v_i^T,$$

$$\implies \mathbf{L}_G^e(\mathbf{x}) \succeq \lambda_2(\mathbf{L}_G^e(\mathbf{x}))(\mathbf{I}_n - v_1 v_1^T),$$

$$\implies \mathbf{L}_G^e(\mathbf{x}) \succeq \lambda_2(\mathbf{L}_G^e(\mathbf{x})) \left(\mathbf{I}_n - \frac{1}{n}\mathbf{1}\mathbf{1}^T\right). \tag{3.11}$$

The above analysis is for any given attacker's strategy $e \subseteq \mathcal{A}$. Next, we show that our modified algebraic connectivity maximization problem is equivalent to $\max_{\mathbf{x},\beta} \beta$ in (3.8), i.e., $\max_{\mathbf{x},\beta} \beta = \lambda_2(\mathbf{L}_G^{e^*}(\mathbf{x}^*))$, where $\mathbf{x}^*$ and e^* are the optimal decisions. For convenience, we denote $\beta^* = \max_{\mathbf{x},\beta} \beta$. The proof includes two parts. First, we show that $\beta^* \geq \lambda_2(\mathbf{L}_G^{e^*}(\mathbf{x}^*))$. We aim to maximize the algebraic connectivity $\lambda_2(\mathbf{L}_G^e(\mathbf{x}))$, and $(\mathbf{x}^*, e^*)$ is a feasible solution pair. Therefore, based on (3.11), $\beta^* \geq \lambda_2(\mathbf{L}_G^e(\mathbf{x}^*))$ should hold. Second, we show that $\beta^* \leq \lambda_2(\mathbf{L}_G^{e^*}(\mathbf{x}^*))$. Since $\beta^*, e^*, \mathbf{x}^*$ are feasible, then, the constraints in (3.8) should be satisfied, i.e., $L_G^e(\mathbf{x}^*) \succeq \beta^*(\mathbf{I}_n - \frac{1}{n}\mathbf{1}\mathbf{1}^T)$, $\forall e \subseteq \mathcal{A}, |e| = \psi$, which gives $L_G^{e^*}(\mathbf{x}^*) \succeq \beta^*(\mathbf{I}_n - \frac{1}{n}\mathbf{1}\mathbf{1}^T)$. Let μ be any unit vector that satisfies $\mu^T v_1 = 0$. Then, we obtain $\mu^T \mathbf{L}_G^{e^*}(\mathbf{x}^*)\mu \geq \mu^T \beta^*(\mathbf{I}_n - \frac{1}{n}\mathbf{11}^T)\mu \rightarrow \mu^T \mathbf{L}_G^{e^*}(\mathbf{x}^*)\mu \geq \beta^*\mu^T \mathbf{I}_n\mu - \beta^*\mu^T v_1 v_1^T \mu \rightarrow \mu^T \mathbf{L}_G^{e^*}(\mathbf{x}^*)\mu \geq \beta^*\mu^T \mathbf{I}_n\mu = \beta^*$. Since vector μ is not fixed, and based on (2.9), we have $\lambda_2(\mathbf{L}_G^{e^*}(\mathbf{x}^*)) \geq \beta^*$. Therefore, $\max_{\mathbf{x},\beta} \beta = \lambda_2(\mathbf{L}_G^{e^*}(\mathbf{x}^*))$ holds, and (3.7) is equivalent to (3.8). $\qquad\square$

Note that the constraints in (3.8) ensure the network operator to achieve the maximum network connectivity by considering all the possible link removal attacks.

Next, we define a new Stackelberg game $\widetilde{\Xi}_\gamma := \{\mathcal{N}_\gamma, \mathcal{X}_\gamma, \mathcal{A}, \alpha_\gamma, \lambda_2\}$, for $\gamma \in \{1, 2\}$, where $\mathcal{N}_\gamma$, $\mathcal{X}_\gamma$ and $\mathcal{A}$ are the same as those defined in game Ξ_γ; α_γ and λ_2 are the objective functions of the network designer and attacker, respectively. Based on (3.4), (3.5) and Theorem 3.3, the network designer γ's problem is formulated as follows, for $\gamma \in \{1, 2\}$:

$$\widetilde{Q}_\gamma^k: \quad \max_{\mathbf{x}_\gamma(k+c_\gamma), \alpha_\gamma(k+c_\gamma)} \quad \alpha_\gamma(k + c_\gamma)$$

$$\text{s.t. } \mathbf{L}_G^e(k + c_\gamma) \succeq \alpha_\gamma(k + c_\gamma)\left(\mathbf{I}_n - \frac{1}{n}\mathbf{11}^T\right),$$

$$\forall e \subseteq \mathcal{A}, \ |e| = \psi,$$

$$2\{x_i(k + c_\gamma) - x_j(k + c_\gamma)\}^T \{x_i(k) - x_j(k)\}$$
$$= Z_{ij}(k + c_\gamma) + Z_{ij}(k), \quad \forall i, j \in V_\gamma,$$

$$\|x_{ij}(k + c_\gamma)\|_2 \geq \rho_\gamma, \quad \forall (i, j) \in E_\gamma,$$

$$\|x_{ij}(k + c_\gamma)\|_2 \geq \rho_{12}, \quad \forall i \in V_\gamma, \ \forall j \in V_{-\gamma},$$

$$\|x_i(k + c_\gamma) - x_i(k)\|_2 \leq d_\gamma, \quad \forall i \in V_\gamma,$$

$$x_j(k + c_\gamma) = x_j(k), \quad \forall j \in V_{-\gamma}.$$

Note that Laplacian matrices $\mathbf{L}_G^e(k + c_\gamma)$, for $\gamma = 1, 2$, are constructed based on (3.5).

The above analysis leads to the following corollary.

Corollary 3.3 *The Stackelberg game $\widetilde{\Xi}_\gamma$ is strategically equivalent to the game Ξ_γ defined in Sect. 3.2.1.3, for $\gamma \in \{1, 2\}$. The interactions between two network operators can be captured by a strategic equivalent Nash game denoted by $\widetilde{\Xi}_I$, where $\widetilde{\Xi}_I$ includes α_γ, $\gamma = 1, 2$.*

3.2.2.2 Meta-Equilibrium Solution Concept

Stackelberg Equilibrium of the Adversarial Game $\widetilde{\Xi}_\gamma$

In the Stackelberg game, the attacker's strategy is the best response to the action that network designer chooses. Recall that $\Lambda(\mathbf{x}_1, \mathbf{x}_2, e) = \lambda_2(\mathbf{L}_G^e(\mathbf{x}))$. Then, following Sect. 2.1.3, the formal definition of best response is as follows.

Definition 3.2 (*Best Response*) For a given strategy pair $(\mathbf{x}_1, \mathbf{x}_2)$, where $\mathbf{x}_1 \in \mathcal{X}_1$ and $\mathbf{x}_2 \in \mathcal{X}_2$, the best response of the attacker is defined by $BR(\mathbf{x}_1, \mathbf{x}_2) := \{e' : \Lambda(\mathbf{x}_1, \mathbf{x}_2, e') \leq \Lambda(\mathbf{x}_1, \mathbf{x}_2, e), \forall e, e' \subseteq \mathcal{A}, |e'| = |e| = \psi\}$.

Thus, we give the definition of the Stackelberg equilibrium of game $\widetilde{\Xi}_\gamma$, for $\gamma \in \{1, 2\}$.

Definition 3.3 (*Stackelberg Equilibrium*) For a given $\mathbf{x}_{-\gamma} \in \mathcal{X}_{-\gamma}$, the profile $(\mathbf{x}_\gamma^*, e^*)$ constitutes a Stackelberg equilibrium of the adversarial game $\widetilde{\Xi}_\gamma$, for $\gamma \in \{1, 2\}$, if the following conditions are satisfied:

1. Attacker's strategy $e^* \subseteq \mathcal{A}$, where $|e^*| = \psi$, is a best response to $(\mathbf{x}_\gamma^*, \mathbf{x}_{-\gamma})$, i.e., $e^* \in BR(\mathbf{x}_\gamma^*, \mathbf{x}_{-\gamma})$.
2. Network designer γ's strategy $\mathbf{x}_\gamma^* \in \mathcal{X}_\gamma$ satisfies

$$\min_{e \in BR(\mathbf{x}_\gamma^*, \mathbf{x}_{-\gamma})} \Lambda(\mathbf{x}_\gamma^*, \mathbf{x}_{-\gamma}, e) = \max_{\mathbf{x}_\gamma \in \mathcal{X}_\gamma} \min_{e \in BR(\mathbf{x}_\gamma, \mathbf{x}_{-\gamma})} \Lambda(\mathbf{x}_\gamma, \mathbf{x}_{-\gamma}, e) \triangleq \Lambda^{\gamma*},$$

where $\Lambda^{\gamma*}$ is the Stackelberg utility of the designer γ.

Nash Equilibrium of the MAS Network Formation Game $\widetilde{\Xi}_I$

After P_1 takes his action at step k, G_1 and G_{12} are reconfigured, where G_{12} is the network between G_1 and G_2. We denote network G_1 and G_{12} at stage k as $G_{1,k}$ and $G_{12,k}$, respectively. For simplicity, we further define $\widetilde{G}_{12,k} := G_{1,k} \cup G_{12,k}$, which is a shorthand notation for the merged network. Then, network G_k can be expressed as $G_k = \widetilde{G}_{12,k} \cup G_{2,k}$. Similarly, after P_2 updates network G_2 at step k, the whole network G_k becomes $G_k = \widetilde{G}_{21,k} \cup G_{1,k}$, where $\widetilde{G}_{21,k} := G_{2,k} \cup G_{12,k}$. Then, similar to Definition 3.1, the formal definition of NE which depends on the *position* of robots is as follows.

Definition 3.4 (*Nash Equilibrium*) The NE solution to game $\widetilde{\Xi}_I$ is a strategy profile $\mathbf{x}^*$, where $\mathbf{x}^* = (\mathbf{x}_1^*, \mathbf{x}_2^*) \in \mathcal{X}$, that satisfies

$$\lambda_2\left(\mathbf{L}_{G_k}(\mathbf{x}_1^*, \mathbf{x}_2^*)\right) \geq \lambda_2\left(\mathbf{L}_{G_k}(\mathbf{x}_1, \mathbf{x}_2^*)\right),$$
$$\lambda_2\left(\mathbf{L}_{G_k}(\mathbf{x}_1^*, \mathbf{x}_2^*)\right) \geq \lambda_2\left(\mathbf{L}_{G_k}(\mathbf{x}_1^*, \mathbf{x}_2)\right),$$

for $\forall \mathbf{x}_1 \in \mathcal{X}_1$ and $\forall \mathbf{x}_2 \in \mathcal{X}_2$, where k denotes the time step.

Note that $\mathbf{L}_{G_k}$ in Definition 3.4 captures the network characteristic under all possible attacks instead of a particular one. At the NE point, no player can individually increase the global network connectivity by reconfiguring their MAS network.

Meta-equilibrium of the Games-in-Games

To design a secure multi-layer MAS, each network operator should take into account the attacker's behavior and the other network operator's strategy. Hence, a holistic equilibrium solution concept of the proposed games-in-games formulation is necessary which is presented as follows.

Definition 3.5 (*Meta-equilibrium*) The meta-equilibrium of the multi-layer MAS network formation game is captured by the profile $(\mathbf{x}_1^*, \mathbf{x}_2^*, e^*)$ which satisfies the following conditions:

1. $(\mathbf{x}_\gamma^*, e^*)$ constitutes a Stackelberg equilibrium of game $\widetilde{\Xi}_\gamma$, for $\gamma = 1, 2$.
2. $\mathbf{x}^* = (\mathbf{x}_1^*, \mathbf{x}_2^*)$ is an NE of game $\widetilde{\Xi}_I$.

Next, our goal is to find a meta-equilibrium by addressing problems $\widetilde{Q}_\gamma^k$, $\gamma = 1, 2$.

3.2.3   SDP-Based Approach and Online Algorithm

Next, we reformulate the network designer's problem as an SDP and design an algorithm to compute the meta-equilibrium of the MAS network formation game.

3.2.3.1   SDP Reformulation

Notice that in $\widetilde{Q}_\gamma^k$, the minimum distance constraints $||x_{ij}(k + c_\gamma)||_2 \geq \rho_\gamma$, $\forall (i, j) \in E_\gamma$, are *nonconvex*. To address this issue, we regard $Z_{ij}(k + c_\gamma)$ as a new decision variable. Based on the definition $Z_{ij}(t) := ||x_{ij}(t)||_2^2$, we have $||x_{ij}(k + c_\gamma)||_2^2 = Z_{ij}(k + c_\gamma)$. Note that the Laplacian matrix $\mathbf{L}_G^e(k + c_\gamma)$ depends linearly on $Z_{ij}(k + c_\gamma)$, $i, j \in V$, based on the relationships (3.5) and (3.6). Then, we solve problems $\widetilde{Q}_\gamma^k$ with respect to unknowns $Z_{ij}(k + c_\gamma)$ and $\mathbf{x}(k + c_\gamma)$ jointly. In this way, $\widetilde{Q}_\gamma^k$ becomes a convex problem. However, due to the coupling between the robots position and the distance vectors, solving $\widetilde{Q}_\gamma^k$ via merely adding new variables yields inconsistency between the obtained solutions $\mathbf{x}(k + c_\gamma)$ and $Z_{ij}(k + c_\gamma)$, $\forall i, j \in V$. To address this issue, we first present the following definition.

Definition 3.6 (*Euclidean Distance Matrix*) Given the positions of a set of n points denoted by $\{x_1, \ldots, x_n\}$, the Euclidean distance matrix representing the points spacing is defined as

$$\mathbf{D} := [d_{ij}]_{i, j=1,\ldots,n}, \quad \text{where } d_{ij} = ||x_i - x_j||_2^2.$$

A critical property of the Euclidean distance matrix is summarized in the following theorem.

Theorem 3.4 ([42]) *A matrix* $\mathbf{D} = [d_{ij}]_{i, j=1,\ldots,n}$ *is an Euclidean distance matrix if and only if*

$$-\mathbf{CDC} \succeq 0, \text{ and } d_{ii} = 0, \ i = 1, \ldots, n, \tag{3.12}$$

where $\mathbf{C} := \mathbf{I}_n - \frac{1}{n}\mathbf{11}^T$.

Note that (3.12) is a necessary and sufficient condition that ensures $\mathbf{D}$ an Euclidean distance matrix. In addition, the inequality and equality in (3.12) are both convex. Therefore, Theorem 3.4 provides an approach to avoid the inconsistency between the robots position and distance vectors when they are treated as independent variables.

In specific, denote $\mathbf{Z} = [Z_{ij}]_{i,j \in V}$, $\mathbf{C} = \mathbf{I}_n - \frac{1}{n}\mathbf{1}\mathbf{1}^T$, and we can further reformulate problems $\tilde{\mathcal{Q}}_\gamma^k$, $\gamma \in \{1, 2\}$, as

$$
\begin{aligned}
\overline{\mathcal{Q}}_\gamma^k : \quad & \max_{\mathbf{x}_\gamma(k+c_\gamma),\mathbf{Z}(k+c_\gamma),\alpha_\gamma(k+c_\gamma)} \quad \alpha_\gamma(k+c_\gamma) \\
\text{s.t.} \quad & \mathbf{L}_G^e(k+c_\gamma) \succeq \alpha_\gamma(k+c_\gamma)\mathbf{C}, \quad \forall e \subseteq \mathcal{A}, \ |e| = \psi, \\
& 2\{x_i(k+c_\gamma) - x_j(k+c_\gamma)\}^T \{x_i(k) - x_j(k)\} \\
& \qquad = Z_{ij}(k+c_\gamma) + Z_{ij}(k), \quad \forall i, j \in V_\gamma, \\
& Z_{ij}(k+c_\gamma) \geq \rho_\gamma^2, \ \forall (i, j) \in E_\gamma, \\
& Z_{ij}(k+c_\gamma) \geq \rho_{12}^2, \quad \forall i \in V_\gamma, \ \forall j \in V_{-\gamma}, \\
& -\mathbf{C}\mathbf{Z}(k+c_\gamma)\mathbf{C} \succeq 0, \ Z_{ii}(k+c_\gamma) = 0, \ i \in V, \\
& ||x_i(k+c_\gamma) - x_i(k)||_2 \leq d_\gamma, \ \forall i \in V_\gamma, \\
& x_j(k+c_\gamma) = x_j(k), \ \forall j \in V_{-\gamma}.
\end{aligned}
\tag{3.13}
$$

Remark 3.3 P_γ controls robots in G_γ and reconfigures the network by solving $\overline{\mathcal{Q}}_\gamma^k$ to obtain the new positions of robots for $\gamma \in \{1, 2\}$. Furthermore, $\overline{\mathcal{Q}}_\gamma^k$ becomes convex and admits an SDP formulation which can be solved efficiently.

3.2.3.2 Online Algorithm

We have obtained the SDP formulations $\overline{\mathcal{Q}}_\gamma^k$, $\gamma = 1, 2$, and next we aim to find the solution that results in a meta-equilibrium MAS configuration. In the MAS formation game, P_1 controls robots in G_1 and reconfigures the network by solving the optimization problem $\overline{\mathcal{Q}}_1^k$ to obtain a new position of each robot. P_2 controls robots in network G_2 in a similar way by solving $\overline{\mathcal{Q}}_2^k$. Note that the players' action at the current step can be seen as a best-response to the network at the previous step by taking the worst-case attack into account. In addition, for given τ_1 and τ_2 in the continuous time space, we can obtain their update frequencies by normalizing them into integers denoted by c_1 and c_2, respectively. Then, P_1 and P_2 reconfigure their robots for every c_1 and c_2 time intervals which can also be interpreted as the frequency of solving $\overline{\mathcal{Q}}_1^k$ and $\overline{\mathcal{Q}}_2^k$, respectively. Since both players maximize the global network connectivity at every update step, then one approach to find the meta-equilibrium solution is to address $\overline{\mathcal{Q}}_1^k$ and $\overline{\mathcal{Q}}_2^k$ iteratively by two players until the yielding MAS possesses the same secure topology, i.e., P_1 and P_2 cannot increase the network connectivity further through reallocating their robots. Note that this algorithm can be implemented in an *online* fashion. For clarity, Algorithm 3.1 shows the updating details.

Algorithm 3.1 Secure and resilient MAS network formation

1: Initialize mobile robots' position $x_i(0)$, $\forall i \in V$, $\mathbf{x}(1) = 2\mathbf{x}(0)$, $k = 1$, $\kappa = 10^{-6}$.
2: **for** $k = 1$ **do**
3:     **while** $k \mod c_1 = 0$ and $\|\mathbf{x}(k) - \mathbf{x}(k - c_1)\|_\infty > \kappa$ **do**
4:         P_1 obtains new strategy $\mathbf{x}_1(k + c_1)$ via solving $\overline{Q}_1^k$
5:     **end while**
6:     **while** $k \mod c_2 = 0$ and $\|\mathbf{x}(k) - \mathbf{x}(k - c_2)\|_\infty > \kappa$ **do**
7:         P_2 obtains new strategy $\mathbf{x}_2(k + c_2)$ via solving $\overline{Q}_2^k$
8:     **end while**
9:     $k \leftarrow k + 1$
10: **end for**
11: **return** $\mathbf{x}(k)$

A typical example of the algorithm is *alternating update* in which P_1 and P_2 have the same update frequency but not update at the same time and reconfigure the MAS network sequentially.

3.2.3.3 Analytical Results

Regarding the feasibility of the problems $\overline{Q}_\gamma^k$ for $\gamma = 1, 2$, we have the following theorem.

Theorem 3.5 *For a given multi-layer MAS network where the distance between robots satisfies the predefined minimum distance constraint, problems $\overline{Q}_1^k$ and $\overline{Q}_2^k$ are always feasible.*

Proof Since problems $\overline{Q}_1^k$ and $\overline{Q}_2^k$ are similar, without loss of generality, we only analyze $\overline{Q}_1^k$. Given an initial configuration of the network, we calculate its algebraic connectivity, and denote it as α_1^0. P_1 aims to maximize the algebraic connectivity by updating his network. For the first step, if the algebraic connectivity α_1^1 of the network is no larger than α_1^0 after P_1 taking his action $\mathbf{x}_1$, i.e., $\alpha_1^0 \geq \alpha_1^1$, then, the optimal solution to $\overline{Q}_1^k$ is α_1^0 which means that P_1 will not modify the configuration of the current network. Therefore, for a given MAS network, a feasible solution to $\overline{Q}_1^k$ always exists.   □

When $\overline{Q}_1^k$ and $\overline{Q}_2^k$ are feasible at each update step, another critical property is the convergence of the proposed iterative algorithm. The result is summarized in Theorem 3.6.

Theorem 3.6 *The proposed online Algorithm 3.1 of the adversarial network forma-tion game converges to a meta-equilibrium asymptotically.*

Proof First, recall that both players maximize the algebraic connectivity of the global network by anticipating the worst-case attack $e^* \subseteq \mathcal{A}$, and thus the result-ing $\alpha_\gamma(k + c_\gamma)$, $\gamma \in \{1, 2\}$, is no less than the one obtained from the previous update step which yields a non-decreasing network connectivity sequence λ_2. In addition, for a network with n nodes, its algebraic connectivity is upper bounded by a value depending on $f(d_\gamma)$ [43]. Thus, based on the monotone convergence theorem [44], we can conclude that the network connectivity sequence converges asymptotically. Denote the actions of two players that achieve the network connectivity limit as $\bar{\mathbf{x}}_1$ and $\bar{\mathbf{x}}_2$ at some step l, and then, we obtain $\lambda_2\big(\mathbf{L}_{G_l}^{e^*}(\bar{\mathbf{x}}_1, \bar{\mathbf{x}}_2)\big) \geq \lambda_2\big(\mathbf{L}_{G_l}^{e^*}(\mathbf{x}_1, \bar{\mathbf{x}}_2)\big)$, $\lambda_2\big(\mathbf{L}_{G_l}^{e^*}(\bar{\mathbf{x}}_1, \bar{\mathbf{x}}_2)\big) \geq \lambda_2\big(\mathbf{L}_{G_l}^{e^*}(\bar{\mathbf{x}}_1, \mathbf{x}_2)\big)$, for $\forall \mathbf{x}_1 \in \mathcal{X}_1$ and $\forall \mathbf{x}_2 \in \mathcal{X}_2$. Otherwise, $\bar{\mathbf{x}}_1$ and $\bar{\mathbf{x}}_2$ do not result in the network connectivity limit. Obviously, the strategy pair $(\bar{\mathbf{x}}_1, \bar{\mathbf{x}}_2)$ satisfies the equilibrium Definition 3.4 which indicates that the proposed iterative algorithm converges to a meta-equilibrium point asymptotically. □

We next investigate the uniqueness of the meta-equilibrium of the game, and the result is shown in the following theorem.

Theorem 3.7 *The meta-equilibrium of the game is not unique, i.e., different equilibrium profiles* $(\mathbf{x}_1^*, \mathbf{x}_2^*, e^*)$ *are possible.*

Proof To show the non-uniqueness of the meta-equilibrium, one possible way is to find a different position pair $(\tilde{\mathbf{x}}_1^*, \tilde{\mathbf{x}}_2^*)$ but the network configuration is the same with a one under the meta-equilibrium, say $(\mathbf{x}_1^*, \mathbf{x}_2^*)$. This can be achieved by the simultaneous offset or rotation in $\mathbf{x}_1^*$ and $\mathbf{x}_2^*$. For example, under the meta-equilibrium $(\mathbf{x}_1^*, \mathbf{x}_2^*, e^*)$, the profile $(\mathbf{x}_1^* + \zeta, \mathbf{x}_2^* + \zeta, e^*)$ is also a meta-equilibrium, where $\zeta \in \mathbb{R}^3$, and this shows the nonuniqueness of the equilibrium. □

Remark 3.4 The network configuration at meta-equilibrium can also be different at which the network exhibits various levels of network connectivity. This phenomenon is further demonstrated through case studies.

Another interesting result is on the effectiveness of our proposed strategy comparing with simpler ones without attack anticipation for the network designers. In the proposed model, if there is no attack, then the network connectivity achieved at the equilibrium is no better than the one designed under without attack considerations. However, when the anticipated attack happens, the designed network through the established framework is no worse than the one without considering adversaries. Characterizing the conditions under which these two classes of strategies coincide is not trivial and it is also related to the system parameters of multi-layer MAS. We next present an illustrative example when these two strategies whether considering attacks are the same.

Example: In a network containing 3 agents with the minimum distance in between equaling 1, the optimal configuration without considering attacks is a regular triangle. In comparison, if the network designer anticipates a link removal, then the best strategy is still to construct a regular triangle network. The reason is that after removing a link, the regular triangle network has the largest connectivity comparing with other networks.

3.2.4 Adversarial Analysis

In this section, we first analyze the security of MAS network by deriving a closed form solution of the jamming attacker, and then present two other types of cyberattacks for further resiliency quantification of the proposed iterative algorithm.

3.2.4.1 Adversarial Analysis

Denote the network as $\widetilde{G}(i, j) = (V, E \backslash (i, j))$ after removing a link $(i, j) \in E$ from network G, then, we have $\widetilde{\mathbf{L}} = \mathbf{L} - \Delta \mathbf{L}$ and $\Delta \mathbf{L} = \Delta \mathbf{D} - \Delta \mathbf{A}$, where $\Delta \mathbf{D}$ and $\Delta \mathbf{A}$ are the decreased degree and adjacency matrices, respectively. By using Eq. (2.7), we obtain $\Delta \mathbf{D}$ and $\Delta \mathbf{A}$ as follows: $\Delta \mathbf{D} = \mathbf{e}_i \tilde{\mathbf{e}}_{i,j}^T + \mathbf{e}_j \tilde{\mathbf{e}}_{j,i}^T$, $\Delta \mathbf{A} = \mathbf{e}_i \tilde{\mathbf{e}}_{j,i}^T + \mathbf{e}_j \tilde{\mathbf{e}}_{i,j}^T$, where $\mathbf{e}_i$ and $\tilde{\mathbf{e}}_{i,j}$ are n-dimensional zero vectors except the ith element equaling to 1 and w_{ij}, respectively, and similar for $\mathbf{e}_j$ and $\tilde{\mathbf{e}}_{j,i}$. Denote the Laplacian matrix of $\widetilde{G}(i, j)$ as $\widetilde{\mathbf{L}}(i, j)$, and by using $\Delta \mathbf{D}$ and $\Delta \mathbf{A}$, we have

$$\widetilde{\mathbf{L}}(i, j) = \mathbf{L} - \left(\mathbf{e}_i - \mathbf{e}_j\right)\left(\tilde{\mathbf{e}}_{i,j} - \tilde{\mathbf{e}}_{j,i}\right)^T. \tag{3.14}$$

When link (i, j) is attacked, the resulting Laplacian is given by (3.14). Denote the Fiedler vector of $\mathbf{L}$ as $\mathbf{u}$, and thus $\mathbf{u}^T \mathbf{L} \mathbf{u} = \lambda_2(\mathbf{L})$ based on the definition. By using Courant–Fisher Theorem in (2.9), we obtain the following:

$$\begin{aligned}
\lambda_2\left(\widetilde{\mathbf{L}}(i, j)\right) &\leq \mathbf{u}^T \widetilde{\mathbf{L}}(i, j) \mathbf{u} \\
&= \mathbf{u}^T \left(\mathbf{L} - \left(\mathbf{e}_i - \mathbf{e}_j\right)\left(\tilde{\mathbf{e}}_{i,j} - \tilde{\mathbf{e}}_{j,i}\right)^T\right)\mathbf{u} \\
&= \mathbf{u}^T \mathbf{L} \mathbf{u} - (u_i - u_j)(w_{ij} u_i - w_{ji} u_j) \\
&= \lambda_2(\mathbf{L}) - w_{ij}(u_i - u_j)^2.
\end{aligned} \tag{3.15}$$

Therefore, by removing the link $(i, j)^*$, where

$$(i, j)^* \in \arg \max_{(i,j) \in E} w_{ij}(u_i - u_j)^2, \tag{3.16}$$

the upper bound of $\lambda_2\left(\widetilde{\mathbf{L}}(i, j)\right)$ is the smallest. The strategy in (3.16) can be seen as a greedy heuristic for the attacker to compromise the network G. Specifically, the attacker can apply the above procedure iteratively to find a set of critical links to accommodate the attacker's ability. To this end, the jamming attacker's strategy is to compromise those links with top ψ largest value of $w_{ij}(u_i - u_j)^2$, $i, j \in V$. Therefore, the network operators designs secure strategies by anticipating that these ψ critical links could be compromised by the attacker.

Remark 3.5 Depending on the scope of knowledge that the attacker has of the network, our proposed framework can be used for attackers of different knowledge levels. For example, an attacker may know the information of the whole multi-layer

MAS network or merely one sub-network. For the former case of attack, closed form solutions have already been presented above. For the latter case, the attacking surface is smaller, and the security analysis can be carried in a similar fashion where an additional constraint on the set of possible compromised links is imposed.

3.2.4.2 GPS Spoofing Cyberattack

We have analyzed the strategic behavior of jamming attacker, and our designed multilayer MAS network is resistant to link removal attacks due to the fact that the network operators anticipate a certain level of attacks when designing strategies. In order to assess the resilience of designed iterative algorithm in Sect. 3.2.3.2, we introduce another type of adversarial attacks to the MAS network called global positioning system (GPS) spoofing attack.

A GPS spoofing attack aims to deceive a GPS receiver in terms of the object's position, velocity and time by generating counterfeit GPS signals [45]. In [46], the authors have demonstrated that UAVs can be controlled by the attackers and go to a wrong position through the GPS spoofing attack. We consider the scenario that the compromised robot is spoofed which can be realized by adding a disruptive position signal to the robot's real control command. Therefore, through the GPS spoofing attack, the mobile robot is controlled by the adversary, but it still maintains communications with other robots in the network. In addition, we assume that the attack cannot last forever but for a period of g_a in the discrete time measure, since the resource of an attacker is limited, and the abnormal/unexpected behavior of the other unattacked robots resulting from the spoofing attack can be detected by the network operator.

Specifically, if robot i, $i \in V$, is compromised by the spoofing attack at time step k_1, and the attack lasts for g_a time steps, then this scenario can be captured by adding the following constraint to $\overline{Q}_\gamma^k$: $x_i(k + 1) = x_i(k) + \epsilon(k)$, $k = k_1, \ldots, k_1 + g_a - 1$, where $\epsilon(k)$ is the disruptive signal added by the attacker. The attacked robot is usually randomly chosen. To evaluate the impact of attack, we choose the robot that has the maximum degree denoted by i_{max} and satisfies

$$i_{max} \in \arg \max_{i \in V} \sum_{j \in \mathcal{N}_i} w_{ij}, \tag{3.17}$$

where $\mathcal{N}_i$ is the set of nodes connected to robot i.

The GPS spoofing attack decreases the network connectivity during the network formation process. The resilience of designed iterative algorithm can be quantified by the increased network performance through the network operators' response to the cyberattacks.

3.2.5   Case Studies

In this section, we use case studies to quantify the security and resiliency of the designed online algorithm, and identify the interdependency in the multi-layer MAS networks. We adopt YALMIP [47] to solve the corresponding SDP problems. Specifically, we consider a two-layer MAS network in which G_1 contains 2 nodes and G_2 contains 6 nodes. To illustrate that the designed framework can be applied to cases where the robots at one layer can further be operated in a decentralized way, we assume that the robots in G_2 are divided into 2 equal-size groups connected by a secure link between nodes 3 and 4. The investigated scenario is applicable when the agents in MAS are sparsely distributed in geometric clusters.

The communication strength between agents follows the one in Fig. 3.5. Further, two layers of MAS are operated in two planes where the third dimension of their position is fixed satisfying the minimum distance ρ_{12}. The initial positions of robots in G_1 (upper layer) are $(1, 3, 1.2)$, $(2, 3, 1.2)$, and robots in G_2 (lower layer) are $(0, 0, 0)$, $(0, 1.5, 0)$, $(1, 0, 0)$, $(2, 0, 0)$, $(3, -1.5, 0)$, $(3, 0, 0)$. The safety distance between robots in G_1 and G_2 is $\rho_1 = \rho_2 = 1$, and the maximum distance that robots at each layer can move at each update step is $d_1 = d_2 = 0.2$. The update frequency of network operator P_1 is two times faster than network operator P_2, i.e., $2c_1 = c_2$. In addition, both network designers prepare for the worst-case single link removal of jamming attack, i.e., $|e| = \psi = 1$, during the MAS network formation.

3.2.5.1   Secure Design of MAS Networks

First, we illustrate the secure design of two-layer MAS network under the jamming attack using Algorithm 3.1. Figure 3.8 shows the results. The final positions of agents in G_1 are $(1.88, -0.13, 1.2)$, $(1.88, 0.87, 1.2)$, and those in G_2 are $(0.94, -0.85, 0)$, $(0.94, 0.65, 0)$, $(1.60, -0.10, 0)$, $(2.26, 0.65, 0)$, $(1.55, -0.29, 0)$, $(2.92, 1.40, 0)$. The connectivity of the integrated MAS network is iteratively improved, and converges to a steady value 1.4 after approximate 40 steps. During the updates, each network operator is aware of the strategic jamming attack that compromises the most critical communication link. Hence, the network shown in Fig. 3.8b is a meta-equilibrium configuration. This example shows the effectiveness of the proposed method in designing secure multi-layer MAS network. To show the nonuniqueness of the equilibrium solutions, we modify the initial positions of agents where the robots in G_1 (upper layer) start with $(3, 1, 1.2)$ and $(3, 2, 1.2)$. The results are shown in Fig. 3.9. We can see that the final positions of agents in G_1 are $(1.06, 0.59, 1.2)$, $(2.06, 0.59, 1.2)$, and those in G_2 are $(0.05, 0.58, 0)$, $(1.55, 1.45, 0)$, $(1.05, 0.58, 0)$, $(2.05, 0.58, 0)$, $(1.55, -0.29, 0)$, $(3.05, 0.58, 0)$. Further, the final network configuration as well as network connectivity are different with the ones shown in Fig. 3.8 which corroborate the meta-equilibrium of the proposed game is not unique.

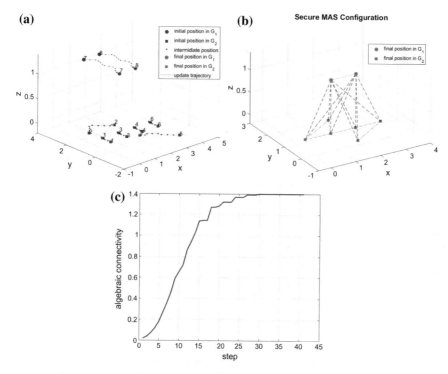

Fig. 3.8 **a** Shows the evolutionary configuration of secure MAS network at each step. **b** depicts the final network configuration. **c** shows the network connectivity under attack with $\psi = 1$

3.2.5.2 Resilience of the Network to Cyberattacks

Second, we investigate the resilience of the designed MAS network to cyberattacks presented in Sect. 3.2.4.2. The metric used for quantifying the resilience is the recovery ability of network connectivity after the adversarial attack.

For the GPS spoofing attack, we assume that it lasts for 5 time steps from step 9–14 before the detection of abnormal movement of MAS by the network operator. Note that the attack duration depends on the detection ability of the network designer. After the identification of attack, the network designer can reboot the compromised agent for it returning to the normal state. Moreover, the horizontal axis in attacker's disruptive command $\epsilon(k), k = 9, \ldots, 14$, is drawn uniformly from $[0, 0.2]$. Figure 3.10 shows the obtained results where agent 7 is compromised. Specifically, the network connectivity encounters a sudden drop at step 9, from 0.58 to 0.37, as shown in Fig. 3.10b due to the spoofing attack. At step 12 which is still in the attacking window, the connectivity, however, has an increase which is a result from the updates of agents at the lower layer G_2. When the spoofing attack is removed, the network recovers quickly after step 14 which shows agile resilience of the proposed control algorithm. Note that the final MAS network configuration is the same as the one in Sect. 3.2.5.1.

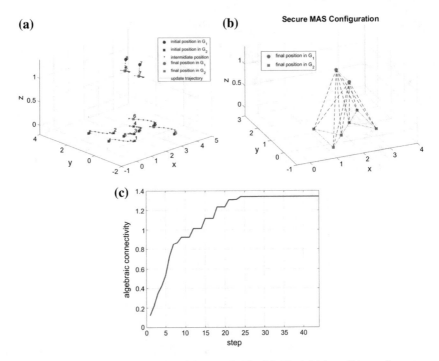

Fig. 3.9 a, b, and **c** Show the results of the ones in Fig. 3.8. The initial conditions of agents are modified and the final equilibrium network is different from the one in Fig. 3.8 which shows the nonuniqueness of the equilibrium

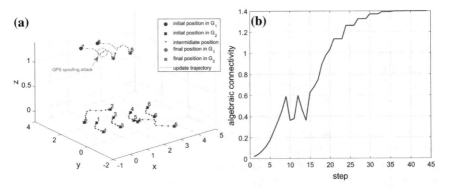

Fig. 3.10 a Shows the evolutionary configuration of secure MAS network at each step. The GPS spoofing attack is introduced at time step 9, and it lasts for 5 steps. The attack duration depends on the detection ability of the network designer. **b** shows the corresponding network connectivity

We next investigate a scenario in which the spoofing attack is introduced after the network reaching an equilibrium. Specifically, the attack launches at step 35 and it lasts for 6 steps. The results are shown in Fig. 3.11. Similar to the previous case, the network can responsd to the attack in a fast fashion and tries to recover

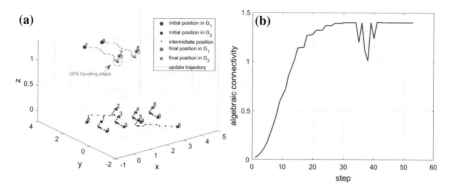

Fig. 3.11 **a** Shows the evolutionary configuration of secure MAS network at each step. **b** shows the corresponding network connectivity. The spoofing attack launches at step 35 and it lasts for 6 steps. The network recovers and reaches a meta-equilibrium quickly after the removal of attack

the network connectivity with its best effort during the attacking window. After the detection and removal of the attack, the network performance is improved and the equilibrium network configuration is achieved which is the same as the previous one before attack.

In summary, the designed two-layer MAS network using Algorithm 3.1 is of high-level situational awareness and is resilient to spoofing cyberattacks.

3.2.5.3 Interdependency in Multi-layer MAS Networks

Finally, we characterize the inherent interdependencies in the two-layer MAS network. The studied scenario is similar to the one in Sect. 3.2.5.1 except that the agents at the lower layer only update once at step 3 and then remain static afterward. This model is applicable to the disaster-response scenario where the constrained movement of agents is caused by physical attacks. The corresponding results are shown in Fig. 3.12. The agents at upper layer move consecutively toward positions that allow to set up the most intra-links with the ones at lower layer, validating the existence of interdependency between two-layer MAS network. Due to the interdependency, the integrated MAS is more resilient to adversarial attacks and natural failures, since the agents at the unattacked layer can response to the emergencies quickly.

3.3 Summary and Notes

In this chapter, we have investigated the static and dynamic network resilience games in which network designers aim to maximize the algebraic connectivity of the global network. For the static network scenario, we have developed alternating plays mech-

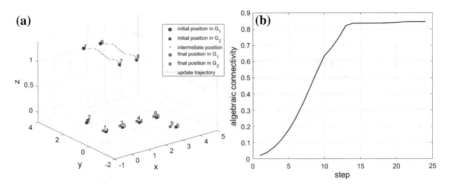

Fig. 3.12 **a** Shows the evolutionary configuration of secure MAS network at each step. Network operator P_2 updates the agents in G_2 only once at time step 3. **b** shows the network connectivity

anism where the players configure or rewire networks step by step, and have shown that the proposed algorithm converges to a Nash equilibrium in a finite number of steps. For the dynamic network scenario, we have established a games-in-games framework for the secure control of multi-layer MAS networks under the adversarial environment. The newly proposed meta-equilibrium solution concept has successfully captured the secure and uncoordinated design of each layer of MAS network through integrative Stackelberg and Nash games. The provided case studies have shown that the designed multi-layer MAS network is of agile resilience to various kinds of cyberattacks.

The readers interested in more details about the network resilience games can refer to [36, 40, 48]. Furthermore, a thorough introduction of network-of-networks and its applications to critical infrastructures is presented in [20]. In addition, for readers interested in cross-layer design, a systematic development of games-in-games principle to achieve robustness, security, and resilience of cyberphysical control systems can be found in [41].

References

1. Zimmerman R, Zhu Q, de Leon F, Guo Z (2017) Conceptual modeling framework to integrate resilient and interdependent infrastructure in extreme weather. J Infrastruct Syst 23(4):04017,034
2. Kurian V, Chen J, Zhu Q (2017) Electric power dependent dynamic tariffs for water distribution systems. In: Proceedings of the 3rd international workshop on cyber-physical systems for smart water networks. ACM, pp 35–38
3. Zimmerman R, Zhu Q, Dimitri C (2018) A network framework for dynamic models of urban food, energy and water systems (fews). Environ Progr Sustain Energy 37(1):122–131
4. Başar T, Olsder GJ (1999) Dynamic noncooperative game theory, 2nd edn. Classics in applied mathematics. SIAM, Philadelphia
5. Fiedler M (1973) Algebraic connectivity of graphs. Czechoslovak Mathematical Journal 23(2):298–305

6. Chen J, Zhou L, Zhu Q (2015) Resilient control design for wind turbines using markov jump linear system model with lévy noise. In: IEEE international conference on smart grid communications (SmartGridComm), pp 828–833

7. Chen J, Zhu Q (2017) A game-theoretic framework for resilient and distributed generation control of renewable energies in microgrids. IEEE Trans Smart Grid 8(1):285–295

8. Zimmerman R, Zhu Q, Dimitri C (2016) Promoting resilience for food, energy, and water interdependencies. J Environ Stud Sci 6(1):50–61

9. Chen J, Zhu Q (2018) A stackelberg game approach for two-level distributed energy management in smart grids. IEEE Trans Smart Grid 9(6):6554–6565

10. Chen J, Zhu Q (2017) Interdependent strategic cyber defense and robust switching control design for wind energy systems. In: IEEE Power & Energy Society General Meeting, pp 1–5

11. Huang L, Chen J, Zhu Q (2017) A large-scale markov game approach to dynamic protection of interdependent infrastructure networks. In: International conference on decision and game theory for security. Springer, pp 357–376

12. Huang L, Chen J, Zhu Q (2017) A factored MDP approach to optimal mechanism design for resilient large-scale interdependent critical infrastructures. In: Workshop on modeling and simulation of cyber-physical energy systems (MSCPES), CPS Week, pp 1–6

13. Huang L, Chen J, Zhu Q (2018) Distributed and optimal resilient planning of large-scale interdependent critical infrastructures. In: Winter simulation conference (WSC), pp 1096–1107

14. Huang L, Chen J, Zhu Q (2018) Factored markov game theory for secure interdependent infrastructure networks. In: Game theory for security and risk management. Springer, Berlin, pp 99–126

15. Ghosh A, Boyd S (2006) Growing well-connected graphs. In: IEEE conference on decision and control, pp 6605–6611

16. Nagarajan H, Rathinam S, Darbha S, Rajagopal K (2012) Algorithms for synthesizing mechanical systems with maximal natural frequencies. Nonlinear Anal Real World Appl 13(5):2154–2162

17. SeDuMi (2015). http://sedumi.ie.lehigh.edu

18. Fisher EB, Neill RP, Ferris MC (2008) Optimal transmission switching. IEEE Trans Power Syst 23(3):1346–1355

19. Simonetto A, Keviczky T, Babuška R (2013) Constrained distributed algebraic connectivity maximization in robotic networks. Automatica 49(5):1348–1357

20. D'Agostino G, Scala A (2014) Networks of networks: the last frontier of complexity, vol 340. Springer, Switzerland

21. Martín-Hernández J, Wang H, Van Mieghem P, D'Agostino G (2014) Algebraic connectivity of interdependent networks. Phys A Stat Mech Appl 404:92–105

22. Chen J, Touati C, Zhu Q (2017) Heterogeneous multi-layer adversarial network design for the IoT-enabled infrastructures. In: IEEE global communications conference, pp 1–6

23. Chen J, Touati C, Zhu Q (2019) Optimal secure two-layer IoT network design. IEEE Trans Control Netw Syst. https://doi.org/10.1109/TCNS.2019.2906893

24. Chen J, Zhu Q (2017) Security as a service for cloud-enabled internet of controlled things under advanced persistent threats: a contract design approach. IEEE Trans Inf Forensics Secur 12(11):2736–2750

25. Chen J, Zhu Q (2016) Optimal contract design under asymmetric information for cloud-enabled internet of controlled things. In: International conference on decision and game theory for security. Springer, Cham, pp 329–348

26. Pawlick J, Chen J, Zhu Q (2019) iSTRICT: an interdependent strategic trust mechanism for the cloud-enabled internet of controlled things. IEEE Trans Inf Forensics Secur 14(6):1654–1669

27. Chen J, Zhu Q (2019) Interdependent strategic security risk management with bounded rationality in the internet of things. IEEE Trans Inf Forensics Secur. https://doi.org/10.1109/TIFS.2019.2911112

28. Xu W, Ma K, Trappe W, Zhang Y (2006) Jamming sensor networks: attack and defense strategies. IEEE Netw 20(3):41–47

29. Dorling K, Heinrichs J, Messier GG, Magierowski S (2017) Vehicle routing problems for drone delivery. IEEE Trans Syst Man Cybern Syst 47(1):70–85
30. Erdelj M, Król M, Natalizio E (2017) Wireless sensor networks and multi-uav systems for natural disaster management. Comput Netw 124:72–86
31. Martin S, Girard A, Fazeli A, Jadbabaie A (2014) Multiagent flocking under general communication rule. IEEE Trans Control Netw Syst 1(2):155–166
32. Olfati-Saber R, Fax JA, Murray RM (2007) Consensus and cooperation in networked multi-agent systems. Proc IEEE 95(1):215–233
33. Nedic A, Ozdaglar A, Parrilo PA (2010) Constrained consensus and optimization in multi-agent networks. IEEE Trans Autom Control 55(4):922–938
34. Chen J, Touati C, Zhu Q (2017) A dynamic game analysis and design of infrastructure network protection and recovery. ACM SIGMETRICS Perform Eval Rev 45(2):125–128
35. Poonawala HA, Satici AC, Eckert H, Spong MW (2015) Collision-free formation control with decentralized connectivity preservation for nonholonomic-wheeled mobile robots. IEEE Trans Control Netw Syst 2(2):122–130
36. Chen J, Zhu Q (2016) Interdependent network formation games with an application to critical infrastructures. In: American control conference (ACC). IEEE, pp 2870–2875
37. Shakeri H, Albin N, Sahneh FD, Poggi-Corradini P, Scoglio C (2016) Maximizing algebraic connectivity in interconnected networks. Phys Rev E 93(3):030,301
38. Tse D, Viswanath P (2005) Fundamentals of wireless communication. Cambridge University Press, Cambridge
39. Chen J, Zhu Q (2019) A games-in-games approach to mosaic command and control design of dynamic network-of-networks for secure and resilient multi-domain operations. In: Proceedings of SPIE. to appear
40. Chen J, Zhu Q (2019) Control of multi-layer mobile autonomous systems in adversarial environments: a games-in-games approach. IEEE Trans Control Netw Syst. submitted
41. Zhu Q, Basar T (2015) Game-theoretic methods for robustness, security, and resilience of cyber-physical control systems: games-in-games principle for optimal cross-layer resilient control systems. IEEE Control Syst Mag 35(1):46–65
42. Dattorro J (2008) Convex optimization & Euclidean distance geometry. Meboo Publishing, USA
43. Godsil C, Royle GF (2013) Algebraic graph theory, vol 207. Springer Science & Business Media, New York
44. Ganter B, Wille R (2012) Formal concept analysis: mathematical foundations. Springer Science & Business Media, Berlin
45. Akos DM (2012) Who's afraid of the spoofer? gps/gnss spoofing detection via automatic gain control (agc). Navigation 59(4):281–290
46. Kerns AJ, Shepard DP, Bhatti JA, Humphreys TE (2014) Unmanned aircraft capture and control via gps spoofing. J Field Robot 31(4):617–636
47. Lofberg J (2004) YALMIP: a toolbox for modeling and optimization in matlab. In: IEEE international symposium on computer aided control systems design, pp 284–289
48. Chen J, Zhu Q (2016) Resilient and decentralized control of multi-level cooperative mobile networks to maintain connectivity under adversarial environment. In: Conference on decision and control (CDC). IEEE, pp 5183–5188

Chapter 4
Interdependent Decision-Making on Complex Networks

4.1 Interdependent Epidemics on Large-Scale Networks

The previous Chap. 3 has focused on the decision-making on finite networks. When the number of agents grows and becomes enormous, e.g., social network and the Internet, the finite network modeling capturing the explicit interactions between agents is inefficient and prohibitive. To this end, this chapter investigates decision-making on complex networks by proposing a new type of system framework.

In a complex network with a large number of agents, we consider the classical susceptible-infected-susceptible (SIS) model in which each agent can be in one of the following two states: susceptible (S) or infected (I). We further consider two strains of interdependent epidemics, strain 1 and strain 2, spreading over the network. Specifically, strains 1 and 2 are in a competing mechanism, i.e., each susceptible agent can either be infected by strain 1 or strain 2 by contacting with other corresponding infected individuals. Let ζ_1 and ζ_2 be the spreading rate of strain 1 and strain 2, respectively. In addition, the infected agents can recover to the susceptible state with rate γ_1 or γ_2 (with respect to strain 1 or strain 2). Besides the self-recovery mechanism, each infected agent can be controlled to return to the healthy state through efforts, e.g., allocating vaccines during flu outbreak season.

To analyze the interdependent epidemic dynamics over complex networks, we consider a degree-based mean-field approximation model [1, 2]. Specifically, the model assumes that the nodes with the same number of degree/connectivity have the identical probability of being infected. Denote by k the degree of a node, where $k \in \mathcal{K} := \{0, 1, 2, \ldots, K\}$, and $P(k) \in [0, 1]$ by the probability distribution of node's degree in the network. Further, we adopt $I_{i,k}(t) \in [0, 1]$ to represent the density of nodes at time t with degree k infected by strain i, $i \in \{1, 2\}$. Then, the dynamics of two competing epidemics can be described by two coupled non-linear differential equations as follows:

© The Author(s), under exclusive license to Springer Nature Switzerland AG 2020
J. Chen and Q. Zhu, *A Game- and Decision-Theoretic Approach to Resilient Interdependent Network Analysis and Design*, SpringerBriefs in Control, Automation and Robotics, https://doi.org/10.1007/978-3-030-23444-7_4

$$\frac{dI_{1,k}(t)}{dt} = -\gamma_1 I_{1,k}(t) + \zeta_1 k[1 - I_{1,k}(t) - I_{2,k}(t)]\Theta_1(t),$$

$$\frac{dI_{2,k}(t)}{dt} = -\gamma_2 I_{2,k}(t) + \zeta_2 k[1 - I_{1,k}(t) - I_{2,k}(t)]\Theta_2(t),$$
(4.1)

where (γ_1, γ_2) and (ζ_1, ζ_2) are the recovery and spreading rates of two strains, respectively. The terms $-\gamma_1 I_{1,k}(t)$ and $-\gamma_2 I_{2,k}(t)$ indicate the proportion of affected nodes returned to the healthy state through recovery. Note that the term $1 - I_{1,k}(t) - I_{2,k}(t)$ captures the density of susceptible nodes with degree k. In addition, $\Theta_i(t)$ represents the probability of a given link connected to a node infected by strain i, and $\Theta_i \in [0, 1]$, $i \in \{1, 2\}$. Specifically, $\Theta_1(t)$ and $\Theta_2(t)$ admit the following expressions:

$$\Theta_1(t) = \frac{\sum_{k' \in \mathcal{K}} k' P(k') I_{1,k'}(t)}{\langle k \rangle},$$
(4.2)

$$\Theta_2(t) = \frac{\sum_{k' \in \mathcal{K}} k' P(k') I_{2,k'}(t)}{\langle k \rangle},$$
(4.3)

where $\langle k \rangle := \sum_k k P(k)$ is the average degree/connectivity of nodes in the network. The nominator $\sum_{k'} k' P(k') I_{i,k'}(t)$ stands for the average connectivity of individuals infected by strain i, $i \in \{1, 2\}$. Note that $\sum_{k'} k' P(k') I_{i,k'}(t) \leq \langle k \rangle$. Therefore, in the class of agents with degree k, the epidemic spreading processes of strain i, $i \in \{1, 2\}$, can be modeled by the term $\zeta_i k[1 - I_{1,k}(t) - I_{2,k}(t)]\Theta_i(t)$ shown in (4.1).

4.2 Controlling Interdependent Epidemics on Complex Networks

With the growth of urban population and the advances in technologies and infrastructures, our world becomes highly connected, and witnesses fast economic development. The connectivity not only enables the communications among mobile and networked devices [3], but also creates dense social and physical interactions in societies, resulting in densely connected complex networks. As the connectivity facilitates the information exchange and interactions [4, 5], its also allows diseases and viruses to spread over the network in multifarious ways. For example, the WannaCry Ransomware has spread through the Internet and infected more than 230,000 computers in over 150 countries. The spreading of Ebola disease in 2014 from West Africa to other countries such as US, UK, and Spain relies on the global connectivity.

Control of the epidemics of diseases and computer viruses is an essential way to mitigate their social and economic impact. Depending on the nature of the epidemics, we can design centralized or distributed policies to contain the growth of the infected population by protecting, removing, and recovering nodes from the population. In human networks where HIN1, HIV, and Ebola viruses can spread, vaccine allocations will be an effective control mechanism. In computer networks that are vulnerable to malware, anti-virus software and quarantine strategies play an essential role in assuring network security.

The control of homogeneous epidemics has found applications in viral marketing [6], computer security [7, 8], and epidemiology [1, 9, 10]. However, with the integration of multiple technologies and the growing complexity of the network systems, homogeneous epidemic models are not sufficient to capture the coexistence of heterogeneous epidemic processes. For example, it has been shown that during the epidemic season, influenza virus can mutate, and during the epidemic season, several types of influenza virus circulate in the human population. An individual cannot be infected by multiple strains simultaneously. Once an individual is infected by one type, he cannot be infected by a virus of a different type. Similarly, in the marketing over social media, two similar products will compete for their customers by spreading information over social networks. An individual who has bought one kind of product is not likely to purchase the same product from another manufacturer. Therefore, it is essential to address the heterogeneous control of interdependent epidemics in a holistic framework.

To this end, the rest of this chapter focuses on the optimal control[1] of two interdependent epidemics spreading over complex networks [11]. To capture the dynamics of the epidemics, we use an SIS epidemic model for both epidemic processes of two strains of viruses 1 and 2, which leads to an epidemic model with three states: (i) susceptible (healthy), (ii) infected by strain 1, and (iii) infected by strain 2. Those infected entities can be treated and moved to the susceptible state through control.

We first study the steady state of the proposed epidemics over complex networks. Through analyzing the non-linear differential equations that model the competing epidemics with control, we observe a non-coexistence phenomenon. Specifically, the network can be in the following three possible equilibrium states: (i) only strain 1, (ii) only strain 2 and (iii) disease-free. Therefore, a coexistence of two competing epidemics in the same network is impossible at the steady state. Furthermore, we investigate the stability of each network equilibrium via the eigenvalue analysis of its linearized dynamic systems.

To design the optimal control strategy, we formulate an optimization problem that minimizes the control cost as well as the severity of epidemics over the network jointly. We propose a gradient-decent algorithm based on a fixed-point iterative scheme to compute the optimal solution and show its convergence to the corresponding fixed-point. In the disease-free regime, we provide a closed-form solution for the optimal control. One critical feature of the policy in this regime is that it is fully determined by the average degree of the epidemic network and the second moment of the degree distribution which yields a distribution independent optimal quarantining strategy. We further observe one emerging phenomenon that, under some conditions, the network encounters a switching of equilibrium through optimal control as the unit cost of effort changes. Depending on the system parameters, the network can be directly controlled to the disease-free equilibrium or from one exclusive equilibrium to the other one first with the symmetric control efforts of two competing epidemics. As long as the applied effort drives the epidemic network to the disease-free equilibrium, the control effort ceases to increase though the unit control cost continues

---

[1]The control effort refers to the applied quarantining strategy.

to decrease. The control effort under which the network switches from exclusive equilibrium of strain 1 or 2 to disease-free regime is referred as *fulfilling threshold*. Finally, we use several numerical experiments on a scale-free network to corroborate the derived theoretical results and discovered phenomenon.

Related Work

The growing social and computer networks provide a fertile medium for the spreading of epidemics. A number of previous works have been on modeling the dynamic processes of epidemics including [1, 12–14]. More recently, a growing number of works have investigated the epidemics spreading on multiplex/interconnected networks [15, 16], and time-varying underlying epidemic networks [17, 18]. Optimal control of single strain epidemics spreading over networks has been considered in various applications including biological disease and virus [19–22], and network security [7, 23, 24]. In addition to the single strain epidemics, the properties of competing epidemics or multi-strain epidemics under different models have been well studied in literature [25–27]. Several methods have been proposed to control multi-strain epidemics over finite networks, including the mean-field approximation-based optimization, impulse control, and passivity-based approach [28–30]. The work in this chapter contributes to the literature on epidemic control by focusing on the control of interdependent epidemics over complex networks. In addition, the established theoretical framework can be further used to address a number of emerging security issues including risk management [31, 32], strategic trust [33, 34], and mechanism design for security [35–37] in cyber-physical systems under adversarial attacks.

4.2.1   Problem Formulation

Based on the model established in Sect. 4.1, the dynamics of two competing epidemics with a control $\mathbf{u} := (u_1, u_2) \in \mathbb{R}_+^2$ can be described as follows:

$$
\begin{aligned}
\frac{dI_{1,k}(t)}{dt} &= -\gamma_1 I_{1,k}(t) + \zeta_1 k[1 - I_{1,k}(t) - I_{2,k}(t)]\Theta_1(t) - u_1 I_{1,k}(t), \\
\frac{dI_{2,k}(t)}{dt} &= -\gamma_2 I_{2,k}(t) + \zeta_2 k[1 - I_{1,k}(t) - I_{2,k}(t)]\Theta_2(t) - u_2 I_{2,k}(t).
\end{aligned}
\tag{4.4}
$$

Recall that (γ_1, γ_2) and (ζ_1, ζ_2) are the recovery and spreading rates of two strains, respectively. The imposed control effort or quarantining strategy to suppress the epidemic spreading is reflected by the terms $-u_1 I_{1,k}(t)$ and $-u_2 I_{2,k}(t)$. Here, based on the mean-field approximation, the agents with different degrees are controlled in the same manner.

The network cost over a time period $[0, T]$ is captured by two terms: the control cost $c_1(\mathbf{u})$, and the severity of epidemics $c_2(w_1 \bar{I}_1(t) + w_2 \bar{I}_2(t))$, where $c_1 : \mathbb{R}_+^2 \to \mathbb{R}_+$, $c_2 : \mathbb{R}_+^2 \to \mathbb{R}_+$ w_1 and w_2 are two positive weighting constants. Note that both c_1 and c_2 are assumed to be continuously differentiable, convex, and monotonically increasing. When $\mathbf{u} = (0, 0)$, we have $c_1(\mathbf{u}) = 0$. If there are no epidemics, then $c_2(0) = 0$. Furthermore, $\bar{I}_1(t)$ and $\bar{I}_2(t)$ are defined as

$$\bar{I}_1(t) := \sum_{k \in \mathcal{K}} P(k) I_{1,k}(t), \tag{4.5}$$

$$\bar{I}_2(t) := \sum_{k \in \mathcal{K}} P(k) I_{2,k}(t), \tag{4.6}$$

respectively, which can be interpreted as the severity of epidemics in the network. The average combined cost of epidemics and control in the long run is given by $\lim_{T \to \infty} \frac{1}{T} \int_0^T c_1(\mathbf{u}) + c_2(w_1 \bar{I}_1(t) + w_2 \bar{I}_2(t)) dt$. Therefore, the average optimal control problem of interdependent epidemics is

$$(\text{OP1}): \quad \min_{\mathbf{u}} \limsup_{T \to \infty} \frac{1}{T} \int_0^T c_1(\mathbf{u}) + c_2 \left(w_1 \bar{I}_1(t) + w_2 \bar{I}_2(t) \right) dt$$

$$\text{s.t.} \quad \text{system dynamics (4.4).}$$

When $\bar{I}_1(t)$ and $\bar{I}_2(t)$ converge to a steady state as $T \to \infty$, the cost functions c_1 and c_2 admit constant values. Therefore, (OP1) can be reformulated as

$$(\text{OP2}): \quad \min_{\mathbf{u}} c_1(\mathbf{u}) + c_2 \left(w_1 \bar{I}_1^*(u_1) + w_2 \bar{I}_2^*(u_2) \right)$$

$$\text{s.t.} \quad \text{system dynamics (4.4),}$$

where $\bar{I}_1^*(u_1)$ and $\bar{I}_2^*(u_2)$ denote the densities of the strains at the steady state under the control $\mathbf{u}$.

To address (OP2), we need to obtain $\bar{I}_1^*(u_1)$ and $\bar{I}_2^*(u_2)$. For convenience, we denote by

$$\psi_i := \frac{\zeta_i}{\gamma_i + u_i}, \quad i = 1, 2, \tag{4.7}$$

the effective spreading rate (ESR) of strains under the control. Note that ESR ψ_i quantifies the net spreading rate of strain i over the network. However, the condition $\psi_i > 1, i \in \{1, 2\}$, alone cannot guarantee the outbreak of the epidemics, as analyzed in Sect. 4.2.2.

At the steady state, $dI_{1,k}/dt = 0$ and $dI_{2,k}/dt = 0$. Then, from (4.4), we obtain

$$I_{1,k} = \frac{\psi_1 k \Theta_1}{1 + \psi_1 k \Theta_1 + \psi_2 k \Theta_2}, \tag{4.8}$$

$$I_{2,k} = \frac{\psi_1 k \Theta_1}{1 + \psi_1 k \Theta_1 + \psi_2 k \Theta_2}. \tag{4.9}$$

Therefore, with (4.8) and (4.9), the optimal control problem (OP2) becomes

$$(\text{OP3}): \quad \min_{\mathbf{u}} \ c_1(\mathbf{u}) + c_2\left(w_1 \bar{I}_1^*(u_1) + w_2 \bar{I}_2^*(u_2)\right)$$

$$\text{s.t.} \quad I_{1,k}^*(u_1) = \frac{\psi_1 k \Theta_1^*}{1 + \psi_1 k \Theta_1^* + \psi_2 k \Theta_2^*}, \quad \forall k \in \mathcal{K},$$

$$I_{2,k}^*(u_2) = \frac{\psi_2 k \Theta_2^*}{1 + \psi_1 k \Theta_1^* + \psi_2 k \Theta_2^*}, \quad \forall k \in \mathcal{K},$$

$$\psi_i = \zeta_i/(\gamma_i + u_i), \quad i = 1, 2,$$

where the variables with superscript $*$ denote the steady state values, i.e., $\Theta_i^* = \frac{\sum_{k'} k' P(k') I_{i,k'}^*(u_i)}{\langle k \rangle}$ and $\bar{I}_i^*(u_i) = \sum_k P(k) I_{i,k}^*(u_i)$, $i \in \{1, 2\}$.

In the suppression of diseases spreading, the control efforts are generally determined by a centralized authority. Thus, our objective is to design a control strategy via solving (OP3) which jointly optimizes the control cost and the epidemics spreading level in the network.

4.2.2   Network Equilibrium and Stability Analysis

To solve the problem (OP3), we first need to analyze the steady states of the epidemics. Substituting (4.8) and (4.9) into (4.2) and (4.3), respectively, yields

$$\Theta_1 = \frac{\psi_1}{\langle k \rangle} \sum_{k' \in \mathcal{K}} \frac{k'^2 P(k') \Theta_1}{1 + \psi_1 k' \Theta_1 + \psi_2 k' \Theta_2}, \tag{4.10}$$

$$\Theta_2 = \frac{\psi_2}{\langle k \rangle} \sum_{k' \in \mathcal{K}} \frac{k'^2 P(k') \Theta_2}{1 + \psi_1 k' \Theta_1 + \psi_2 k' \Theta_2}. \tag{4.11}$$

Thus, the steady state pair (Θ_1^*, Θ_2^*) in (OP3) should satisfy Eqs. (4.10) and (4.11). For clarity, we denote

$$T_1 = \frac{\psi_1 \langle k^2 \rangle}{\langle k \rangle}, \quad T_2 = \frac{\psi_2 \langle k^2 \rangle}{\langle k \rangle}. \tag{4.12}$$

In general, the ESR for different strains of epidemics are unequal, i.e., $\psi_1 \neq \psi_2$. In the special case of $\psi_1 = \psi_2$, the characteristics of two strains are the same, and it can be seen as a generalized single-strain scenario. Therefore, in the following study, we analyze the network equilibrium in the nontrivial regime $\psi_1 \neq \psi_2$.

4.2.2.1   Equilibrium Analysis

For the self-consistency Eqs. (4.10) and (4.11), $(\Theta_1, \Theta_2) = (0, 0)$ is a trivial solution. In this case, $\bar{I}_1^* = \bar{I}_2^* = 0$ which is a disease-free equilibrium. To obtain nontrivial solutions to (4.10) and (4.11), we first present the following theorem.

Theorem 4.1 *There exist no positive solutions to the Eqs.* (4.10) *and* (4.11), *i.e.,* $\Theta_1 > 0$ *and* $\Theta_2 > 0$.

Proof We proof by contradiction. If there exist positive solutions, i.e., $\Theta_1 > 0$ and $\Theta_2 > 0$, (4.10) and (4.11) are equivalent to

$$1 = \frac{\psi_1}{\langle k \rangle} \sum_{k' \in \mathcal{K}} \frac{k'^2 P(k')}{1 + \psi_1 k' \Theta_1 + \psi_2 k' \Theta_2}, \tag{4.13}$$

$$1 = \frac{\psi_2}{\langle k \rangle} \sum_{k' \in \mathcal{K}} \frac{k'^2 P(k')}{1 + \psi_1 k' \Theta_1 + \psi_2 k' \Theta_2}. \tag{4.14}$$

Since $\psi_1 \neq \psi_2$ and $\frac{1}{\langle k \rangle} \sum_{k'} \frac{k'^2 P(k')}{1 + \psi_1 k' \Theta_1 + \psi_2 k' \Theta_2} > 0$, (4.13) and (4.14) cannot be satisfied simultaneously which rules out the positive solutions to Eqs. (4.10) and (4.11). $\quad\square$

Remark 4.1 Based on Theorem 4.1, Θ_1 and Θ_2 cannot be both positive at the steady state, resulting in a *non-coexistence phenomenon* of the two interdependent strains.

The following corollary on the possible nontrivial solutions of Θ_1 and Θ_2 naturally follows from Theorem 4.1.

Corollary 4.1 *The possible nontrivial solutions to* (4.10) *and* (4.11) *fall into two categories: (i)* $\Theta_1 > 0, \Theta_2 = 0$ *and (ii)* $\Theta_2 > 0, \Theta_1 = 0$.

Proof Since $0 \leq \Theta_i \leq 1$, $i = 1, 2$, no negative solutions exist. Then, the possible nontrivial solutions are $\Theta_1 > 0, \Theta_2 = 0$ and $\Theta_2 > 0, \Theta_1 = 0$. $\quad\square$

Corollary 4.1 indicates that, for the possible nontrivial solutions, strain 1 or strain 2 has an exclusive equilibrium. The existence of nontrivial solutions is critical for the analysis of network equilibrium. Therefore, we next investigate the conditions under which the network stabilizes at the exclusive equilibrium.

Theorem 4.2 *Strain i has an exclusive equilibrium if and only if $T_i > 1$, $i \in \{1, 2\}$.*

Proof For the two exclusive equilibria, i.e., $\Theta_1 > 0, \Theta_2 = 0$ and $\Theta_2 > 0, \Theta_1 = 0$, (4.10) and (4.11) are reduced to

$$1 = \frac{\psi_i}{\langle k \rangle} \sum_{k'} \frac{k'^2 P(k')}{1 + \psi_i k' \Theta_i}, \quad i = 1, 2. \tag{4.15}$$

For the former case $\Theta_1 > 0, \Theta_2 = 0$, we define function $g : [0, 1] \rightarrow \mathbb{R}_+$, i.e., $g(\Theta_1) = \frac{\psi_1}{\langle k \rangle} \sum_{k'} \frac{k'^2 P(k')}{1 + \psi_1 k' \Theta_1}$. Then, we obtain

$$g(1) = \frac{\psi_1}{\langle k \rangle} \sum_{k'} \frac{k'^2 P(k')}{1 + \psi_1 k'} = \frac{1}{\langle k \rangle} \sum_{k'} \frac{\psi_1 k'}{1 + \psi_1 k'} k' P(k')$$

$$< \frac{1}{\langle k \rangle} \sum_{k'} k' P(k') = \frac{\langle k \rangle}{\langle k \rangle} = 1.$$

Moreover, $g'(\Theta_1) = -\frac{\psi_1^2}{\langle k \rangle} \sum_{k'} \frac{k'^3 P(k')}{(1+\psi_1 k' \Theta_1)^2} < 0$. Therefore, g is a decreasing function over the domain $\Theta_1 \in [0, 1]$. To ensure the existence of nontrivial solutions to Eq. (4.15), a necessary and sufficient condition is $g(0) > 1$. Since $g(0) = \frac{\psi_1}{\langle k \rangle} \sum_{k'} k'^2 P(k') = \frac{\psi_1 \langle k^2 \rangle}{\langle k \rangle} = T_1$, $g(0) > 1$ is equivalent to $T_1 > 1$. The analysis is similar for the case $\Theta_2 > 0$, $\Theta_1 = 0$, and the necessary and sufficient condition is $T_2 > 1$.   □

Three possible equilibria are summarized as follows:

(1) Disease-free equilibrium, $E_1 = (1, 0, 0)$.
(2) Exclusive equilibrium of strain 1, $E_2 = (\bar{S}_1^*, \bar{I}_1^*, 0)$, if and only if $T_1 > 1$.
(3) Exclusive equilibrium of strain 2, $E_3 = (\bar{S}_2^*, 0, \bar{I}_2^*)$, if and only if $T_2 > 1$.

Remark 4.2 $T_i > 1$ is equivalent to $\psi_i > \frac{\langle k \rangle}{\langle k^2 \rangle}$, $i = 1, 2$. In addition, strain i dies out when ψ_i does not satisfy the condition, and the steady state of the network is the disease-free equilibrium E_1.

4.2.2.2 Stability Analysis of Equilibria

We next analyze the stability of the candidate equilibria presented in Sect. 4.2.2.1.

For convenience, we define $S := \{k \in \mathbb{Z}_+ | P(k) > 0\}$, with $d = |S|$ denoting the Cardinality of S. Then, we have the following result for disease-free equilibrium.

Theorem 4.3 *The disease-free equilibrium E_1 is asymptotically stable if and only if $T_1 \leq 1$ and $T_2 \leq 1$.*

Proof The Jacobian matrix $\mathbf{J}_1 \in \mathbb{R}^{2d \times 2d}$ of the nonlinear dynamic systems (4.4) at the disease-free equilibrium has the block structure $\mathbf{J}_1 = \begin{bmatrix} \mathbf{E}_1 & 0 \\ 0 & \mathbf{E}_2 \end{bmatrix}$, where $\mathbf{E}_i \in \mathbb{R}^d \times \mathbb{R}^d$, $i = 1, 2$. Specifically, $(\mathbf{E}_i)_{k,k'} = \frac{\partial}{\partial I_{i,k'}(t)} \left(\frac{dI_{i,k}(t)}{dt} \right) = \zeta_i k[1 - I_{1,k}(t) - I_{2,k}(t)] \frac{k'P(k')}{\langle k \rangle} = \frac{\zeta_i kk'P(k')}{\langle k \rangle}$, $k' \neq k \in S$, $i = 1, 2$. In addition, $(\mathbf{E}_i)_{k,k} = \frac{\partial}{\partial I_{i,k}(t)} \left(\frac{dI_{i,k}(t)}{dt} \right) = -(\gamma_i + u_i) - \zeta_i k \Theta_i(t) + \zeta_i k[1 - I_{1,k}(t) - I_{2,k}(t)] \frac{kP(k)}{\langle k \rangle} = -(\gamma_i + u_i) + \frac{\zeta_i k^2 P(k)}{\langle k \rangle}$, $k \in S$, $i = 1, 2$. Moreover, for $\forall k, k' \in S$, we have

$$(\mathbf{J}_1)_{k,(d+k')} = \frac{\partial}{\partial I_{2,k'}(t)} \left(\frac{dI_{1,k}(t)}{dt} \right) = -\zeta_1 k \Theta_1(t) = 0,$$

$$(\mathbf{J}_1)_{(d+k),k'} = \frac{\partial}{\partial I_{1,k'}(t)} \left(\frac{dI_{2,k}(t)}{dt} \right) = -\zeta_2 k \Theta_2(t) = 0.$$

Therefore, the compact forms can be expressed as

$$(\mathbf{E}_1)_{k,k'} = \frac{\zeta_1 kk'P(k')}{\langle k \rangle} - (\gamma_1 + u_1)\Delta_{k,k'}, \qquad (4.16)$$

$$(\mathbf{E}_2)_{k,k'} = \frac{\zeta_2 kk'P(k')}{\langle k \rangle} - (\gamma_2 + u_2)\Delta_{k,k'}, \qquad (4.17)$$

where $k, k' \in \mathcal{S}$, and $\Delta_{k,k'} = \begin{cases} 0, & \text{if } k \neq k', \\ 1, & \text{if } k = k'. \end{cases}$

The eigenvalues of the Jacobian matrix $\mathbf{J}_1$ are the union of eigenvalues of the block matrices $\mathbf{E}_1$ and $\mathbf{E}_2$. Therefore, by calculating the determinants $\det(\lambda_1 \mathbf{I}_d - \mathbf{E}_1)$ and $\det(\lambda_2 \mathbf{I}_d - \mathbf{E}_2)$, and setting them to 0, i.e., $\det(\lambda_1 \mathbf{I}_d - \mathbf{E}_1) = 0$ and $\det(\lambda_2 \mathbf{I}_d - \mathbf{E}_2) = 0$, where $\mathbf{I}_d$ is a d-dimensional identity matrix; λ_1 and λ_2 are the eigenvalues to $\mathbf{E}_1$ and $\mathbf{E}_2$, respectively, we obtain the characteristic polynomials

$$(\lambda_1 + \gamma_1 + u_1)^{d-1}\left(\lambda_1 + \gamma_1 + u_1 - \zeta_1 \frac{\langle k^2 \rangle}{\langle k \rangle}\right) = 0, \tag{4.18}$$

$$(\lambda_2 + \gamma_2 + u_2)^{d-1}\left(\lambda_2 + \gamma_2 + u_2 - \zeta_2 \frac{\langle k^2 \rangle}{\langle k \rangle}\right) = 0. \tag{4.19}$$

Therefore, $\mathbf{J}_1$ has eigenvalues $(-\gamma_i - u_i)$ with multiplicity $d - 1$, and $(\zeta_i \frac{\langle k^2 \rangle}{\langle k \rangle} - \gamma_i - u_i)$ with multiplicity 1, $i = 1, 2$. If the disease-free equilibrium $\mathbf{E}_1$ is asymptotically stable, the real parts of eigenvalues of $\mathbf{J}_1$ should be strictly smaller than 0. Obviously, $-\gamma_1 - u_1 < 0$, and $-\gamma_2 - u_2 < 0$. Hence, we further need $\zeta_i \frac{\langle k^2 \rangle}{\langle k \rangle} - \gamma_i - u_i < 0$ which leads to $\frac{\zeta_i}{\gamma_i + u_i} \frac{\langle k^2 \rangle}{\langle k \rangle} < 1 \implies T_i < 1, \ i = 1, 2.$ $\qquad\square$

We further investigate the stability of exclusive equilibrium of strain 1, and the result is presented as follows.

Theorem 4.4 *The exclusive equilibrium of strain 1, E_2, is asymptotically stable if and only if $T_1 > 1$ and $T_1 > T_2$.*

Proof For the exclusive equilibrium of strain 1, note that $\Theta_2(t) = 0$ and thus $I_{2,k}(t) = 0$ when $t \to \infty, k \in \mathcal{S}$. We linearize the system of dynamic equations (4.4) around the equilibrium point E_2, and obtain its Jacobian matrix $\mathbf{J}_2$ which has the following block structure $J_2 = \begin{bmatrix} \mathbf{Z}_1 & \mathbf{Z}_2 \\ 0 & \mathbf{Z}_3 \end{bmatrix}$, where $\mathbf{Z}_j \in \mathbb{R}^d \times \mathbb{R}^d, \ j \in \{1, 2, 3\}$. More specifically,

$(\mathbf{Z}_1)_{k,k'} = \frac{\partial}{\partial I_{1,k'}(t)}\left(\frac{dI_{1,k}(t)}{dt}\right) = \zeta_1 k[1 - I_{1,k}(t) - I_{2,k}(t)]\frac{k'P(k')}{\langle k \rangle} = \frac{\zeta_1 kk'[1 - I_{1,k}(t)]P(k')}{\langle k \rangle},$

$k' \neq k \in \mathcal{S}$, and $(\mathbf{Z}_1)_{k,k} = \frac{\partial}{\partial I_{1,k}(t)}\left(\frac{dI_{1,k}(t)}{dt}\right) = -(\gamma_1 + u_1) - \zeta_1 k \Theta_1(t) + \zeta_1 k[1 - I_{1,k}$

$(t) - I_{2,k}(t)]\frac{kP(k)}{\langle k \rangle} = -(\gamma_1 + u_1) - \zeta_1 k \Theta_1(t) + \frac{\zeta_1 k^2[1 - I_{1,k}(t)]P(k)}{\langle k \rangle},$ for $k \in \mathcal{S}$. For the

block $\mathbf{Z}_2$, we have $(\mathbf{Z}_2)_{k,k'} = \frac{\partial}{\partial I_{2,k}(t)}\left(\frac{dI_{1,k}(t)}{dt}\right) = -\zeta_1 k \Theta_1(t), \ k, k' \in \mathcal{S}$. For the block

$\mathbf{Z}_3$, we have $(\mathbf{Z}_3)_{k,k'} = \frac{\partial}{\partial I_{2,k'}(t)}\left(\frac{dI_{2,k}(t)}{dt}\right) = \zeta_2 k[1 - I_{1,k}(t) - I_{2,k}(t)]\frac{k'P(k')}{\langle k \rangle} =$

$\frac{\zeta_2 kk'[1 - I_{1,k}(t)]P(k')}{\langle k \rangle}, \ k' \neq k \in \mathcal{S}$. In addition, $(\mathbf{Z}_3)_{k,k} = \frac{\partial}{\partial I_{2,k}(t)}\left(\frac{dI_{2,k}(t)}{dt}\right) = -(\gamma_2 + u_2)$

$- \zeta_2 k \Theta_2(t) + \zeta_2 k[1 - I_{1,k}(t) - I_{2,k}(t)]\frac{kP(k)}{\langle k \rangle} = -(\gamma_2 + u_2) + \frac{\zeta_2 k^2[1 - I_{1,k}(t)]P(k)}{\langle k \rangle}, \ k \in$

$\mathcal{S}$. Moreover, for $\forall k, k' \in \mathcal{S}$,

$$(\mathbf{J}_2)_{(d+k),k'} = \frac{\partial}{\partial I_{1,k'}(t)}\left(\frac{dI_{2,k}(t)}{dt}\right) = -\zeta_2 k \Theta_2(t) = 0.$$

Therefore, $\mathbf{J}_2$ admits an upper triangular structure. To analyze its eigenvalues, we only need to compute the eigenvalues of matrices $\mathbf{Z}_1$ and $\mathbf{Z}_3$. For $\mathbf{Z}_1$, finding its determinant and setting it to 0, i.e., $\det(\lambda_3 \mathbf{I}_d - \mathbf{Z}_1) = 0$, lead to

$$\sum_k \frac{kP(k)(I_{1,k}^* - 1)}{\gamma_1 + u_1 + \lambda_3 + \zeta_1 k\Theta_1^*} + \langle k \rangle \frac{\gamma_1 + u_1}{\zeta_1} = 0, \tag{4.20}$$

where λ_3 is the eigenvalue of $\mathbf{Z}_1$. From (4.8) and by using $\Theta_2^* = 0$, we obtain $I_{1,k}^* + I_{1,k}^* \psi_1 k\Theta_1^* = \psi_1 k\Theta_1^*$, which is equivalent to $(\gamma_1 + u_1)I_{1,k}^* + I_{1,k}^* \zeta_1 k\Theta_1^* = \zeta_1 k\Theta_1^*$ and further yields

$$I_{1,k}^* - 1 = \frac{(\gamma_1 + u_1)I_{1,k}^*}{\zeta_1 k\Theta_1^*}. \tag{4.21}$$

Plugging (4.21) into (4.20) yields $\sum_k \frac{kP(k)I_{1,k}^*}{\gamma_1 + u_1 + \lambda_3 + \zeta_1 k\Theta_1^*} - \sum_k \frac{kP(k)I_{1,k}^*}{\gamma_1 + u_1 + \lambda_3} = 0$. Therefore, $\lambda_3 = -\zeta_1 k\Theta_1^* < 0$, $\forall k \in \mathcal{S}$.

For block $\mathbf{Z}_3$, finding its determinant and setting it to 0, i.e., $\det(\lambda_4 \mathbf{I}_d - \mathbf{Z}_3) = 0$, yields $(\lambda_4 + \gamma_2 + u_2)^{d-1}\left(\frac{\psi_2}{\langle k \rangle}\sum_k \frac{kP(k)(1 - I_{1,k}^*)}{\gamma_1 + u_1 + \lambda_4} - 1\right) = 0$. Therefore, $\mathbf{Z}_3$ has negative eigenvalue $-\gamma_2 - u_2$ with multiplicity $d - 1$. Moreover, $\frac{\psi_2}{\langle k \rangle}\sum_k \frac{kP(k)(1 - I_{1,k}^*)}{\gamma_1 + u_1 + \lambda_4} - 1 = 0$ yields $\lambda_4 = \frac{\langle k \rangle}{\langle k^2 \rangle}(\gamma_2 + u_2)(\frac{\zeta_1}{\gamma_1 + u_1} - \frac{\zeta_2}{\gamma_2 + u_2})$. To ensure $\lambda_4 < 0$, we need $\zeta_1/(\gamma_1 + u_1) - \zeta_2/(\gamma_2 + u_2) < 0$, which is equivalent to $T_1 > T_2$.

Hence, $T_1 > T_2$ together with $T_1 > 1$ in Theorem 4.2 lead to the asymptotically stable exclusive equilibrium of strain 1. $\square$

Similarly, we can obtain the condition for stable exclusive equilibrium E_3 as follows.

Theorem 4.5 *The exclusive equilibrium of strain 2, E_3, is asymptotically stable if and only if $T_2 > 1$ and $T_2 > T_1$.*

Proof The proof is similar to that in Theorem 4.5 and hence omitted here. $\square$

In Theorems 4.3, 4.4, and 4.5, the ESR plays an critical role in determining the equilibrium. For example, if ESR of both strains of epidemics are smaller than $\frac{\langle k \rangle}{\langle k^2 \rangle}$, then both epidemics die out at steady state. This disease-free stable state occurs when either the control effort is sufficiently large or the epidemics have a relatively low spreading ability. In comparison, when strain 1's ESR exceeds $\frac{\langle k \rangle}{\langle k^2 \rangle}$ and it is also greater than strain 2's ESR, then only strain 1 exists at equilibrium as shown in Theorem 4.4. This non-coexistence phenomenon indicates that the strain that has a larger spreading rate and is more loosely controlled can eventually survive in the network.

4.2.3 *Optimal Quarantining Strategy Design*

We have obtained the stable equilibria of the interdependent epidemics in Sect. 4.2.2 which further characterize the steady state expressions of parameters in (OP3). In this section, we aim to determine the optimal quarantining strategy of epidemics spreading via solving (OP3) in Sect. 4.2.1.

4.2.3.1 Bounds on Control Effort

Before addressing (OP3), we present the control bounds at each network equilibrium which should be taken into account when designing the optimal control. The following Corollary 4.2 directly follows from Theorems 4.3, 4.4, and 4.5.

Corollary 4.2 *The control efforts leading to different network equilibria are summarized as follows.*

1. *If the network reaches the disease-free equilibrium E_1, the control law needs to satisfy*

$$u_1 \geq \frac{\zeta_1 \langle k^2 \rangle}{\langle k \rangle} - \gamma_1, \tag{4.22}$$

$$u_2 \geq \frac{\zeta_2 \langle k^2 \rangle}{\langle k \rangle} - \gamma_2. \tag{4.23}$$

Note that $u_i \geq 0, i = 1, 2$, and thus when $\frac{\zeta_i \langle k^2 \rangle}{\langle k \rangle} - \gamma_i \leq 0, i = 1, 2$, (4.22) and (4.23) hold.

2. *If the network is stabilized at the exclusive equilibrium E_2, the control law needs to satisfy*

$$u_1 < \frac{\zeta_1 \langle k^2 \rangle}{\langle k \rangle} - \gamma_1, \tag{4.24}$$

$$u_2 > \frac{\zeta_2 (\gamma_1 + u_1)}{\zeta_1} - \gamma_2. \tag{4.25}$$

3. *If the network is stabilized at the exclusive equilibrium E_3, the control law needs to satisfy*

$$u_2 < \frac{\zeta_2 \langle k^2 \rangle}{\langle k \rangle} - \gamma_2, \tag{4.26}$$

$$u_1 > \frac{\zeta_1 (\gamma_2 + u_2)}{\zeta_2} - \gamma_1. \tag{4.27}$$

The control bounds presented in Corollary 4.2 have natural interpretations. The efforts to control strains 1 and 2 by the network operator need to be higher than the thresholds shown in (4.22) and (4.23) to achieve a disease-free steady state. In comparison, if only one strain of epidemics exists at the equilibrium, then the control effort to the other strain is upper bounded by a constant as shown in (4.24) and (4.26).

4.2.3.2 Optimal Quarantine of Interdependent Epidemics

In this section, we address the optimal control problem for each equilibrium case.
Stable disease-free equilibrium
In this case, the optimization problem (OP3) is reduced to

$$(\text{OP4}) : \quad \min_{\mathbf{u}} \quad c_1(\mathbf{u}) + c_2(0)$$

$$\text{s.t. inequalities (4.22) and (4.23).}$$

Due to the monotonicity of function c_1, we can obtain the optimal control solutions based on Corollary 4.2 as

$$u_1 = \max\left(0, \frac{\zeta_1 \langle k^2 \rangle}{\langle k \rangle} - \gamma_1\right),$$

$$u_2 = \max\left(0, \frac{\zeta_2 \langle k^2 \rangle}{\langle k \rangle} - \gamma_2\right). \tag{4.28}$$

When $\frac{\zeta_1 \langle k^2 \rangle}{\langle k \rangle} < \gamma_1$ and $\frac{\zeta_2 \langle k^2 \rangle}{\langle k \rangle} < \gamma_2$, then no control is required and the network reaches the disease-free equilibrium automatically at the steady state due to sufficiently high recovery rates γ_1 and γ_2 of the epidemics comparing with their spreading rates ζ_1 and ζ_2. We summarize the results of optimal quarantine at disease-free regime in the following corollary.

Corollary 4.3 *At the stable disease-free equilibrium, when $\frac{\zeta_1 \langle k^2 \rangle}{\langle k \rangle} < \gamma_1$ and $\frac{\zeta_2 \langle k^2 \rangle}{\langle k \rangle} < \gamma_2$, the optimal effort is irrelevant with network structure, i.e., the degree distribution $P(k)$, and admits a value 0. When $\frac{\zeta_1 \langle k^2 \rangle}{\langle k \rangle} > \gamma_1$ or $\frac{\zeta_2 \langle k^2 \rangle}{\langle k \rangle} > \gamma_2$, the optimal effort is positive and depends on the average network connectivity $\langle k \rangle$ and the second moment $\langle k^2 \rangle$.*

Remark 4.3 In the disease-free regime, Corollary 4.3 indicates that the optimal quarantining strategies for networks with different degree distributions $P(k)$ but the same $\langle k \rangle$ and $\langle k^2 \rangle$ are identical, yielding a *distribution independent* optimal control strategy.

Stable Exclusive Equilibrium of Strain 1
Since $\bar{I}_{2,k}^* = 0$ in this case, the optimization problem (OP3) becomes

$$(\text{OP5}): \quad \min_{\mathbf{u}} \quad c_1(\mathbf{u}) + c_2\left(w_1 \bar{I}_1^*(u_1)\right)$$

$$\text{s.t.} \quad I_{1,k}^*(u_1) = \frac{\psi_1 k \Theta_1^*}{1 + \psi_1 k \Theta_1^*}, \quad \forall k \in \mathcal{K},$$

$$\psi_1 = \zeta_1/(\gamma_1 + u_1),$$

$$\text{inequalities (4.22) and (4.23)},$$

where Θ_1^* and $\bar{I}_1^*(u_1)$ are defined in (OP3).

To solve (OP5), we obtain an expression of $\bar{I}_1^*(u_1)$ with respect to u_1. Note that $I_{1,k}^*(u_1)$, $k \in \mathcal{K}$, and Θ_1^* are coupled in the constraints, and we need to solve the following system of equations:

$$I_{1,k}^*(u_1) = \frac{\psi_1 k \Theta_1^*}{1 + \psi_1 k \Theta_1^*}, \quad k \in \mathcal{K}, \tag{4.29}$$

$$\Theta_1^* = \frac{\sum_{k'} k' P(k') I_{1,k'}^*(u_1)}{\langle k \rangle}. \tag{4.30}$$

To address this problem, we substitute (4.29) into (4.30) and arrive at the following fixed-point equation:

$$\Theta_1^* = \frac{1}{\langle k \rangle} \sum_{k'} \frac{k'^2 P(k') \psi_1 \Theta_1^*}{1 + \psi_1 k' \Theta_1^*}. \tag{4.31}$$

For the existence and uniqueness of the solutions to (4.31), we have the following theorem.

Theorem 4.6 *There exists a unique solution Θ_1^* to the fixed-point Eq. (4.31).*

Proof From the proof of Theorem 4.2, we know that function $g(\Theta_1) = \frac{\psi_1}{\langle k \rangle} \sum_{k'} \frac{k'^2 P(k')}{1 + \psi_1 k' \Theta_1}$ is monotonously decreasing over the domain $\Theta_1 \in [0, 1]$. Moreover, $g(0) > 1$ and $g(1) < 1$. Therefore, $g(\Theta_1) = 1$ has a solution over $\Theta_1 \in [0, 1]$, and the solution is unique. $\qquad \square$

Remark 4.4 The existence and uniqueness of Θ_1^* ensures the *predictability* of $I_{1,k}^*(u_1)$ through (4.29).

Another critical aspect of (OP5) is the continuity of $\bar{I}_1^*(u_1)$ with respect to u_1. When $\bar{I}_1^*(u_1)$ is continuous with u_1, the objective function in (OP5) is a continuous convex function, and thus can be theoretically solved by using the first-order optimality condition directly. When $\bar{I}_1^*(u_1)$ encounters jumps at some points of u_1, which is a possible case, (OP5) becomes challenging to solve, since $c_2(w_1 \bar{I}_1^*(u_1))$ is discontinuous in u_1. If this possible discontinuity feature is neglected, the obtained optimal control law is incorrect. To rule out the probability of discontinuity case, we have the following theorem.

Theorem 4.7 *In the optimization problem (OP5), the mapping $\bar{I}_1^*(u_1)$ is continuous in u_1.*

Proof We need the following lemma for the proof of the theorem.

Lemma 4.1 *Define $H(x, y) := \frac{1}{\langle k \rangle} \sum_{k'} \frac{k'^2 P(k')xy}{1+k'xy}$, where $x \in \mathbb{R}_+$ and $y \in \mathcal{Y} := (0, 1)$. Then, $H(x, y)$ is a contraction in y uniformly over all possible x with contraction constant $0 < c < 1$, i.e., $|H(x, y_1) - H(x, y_2)| \leq c \cdot |y_1 - y_2|$, where $y_1, y_2 \in \mathcal{Y}$.*

Proof First, since $H(x, y)$ is a continuous and twice differentiable function over y, we have

$$\frac{\partial}{\partial y} H(x, y) = \frac{1}{\langle k \rangle} \sum_{k'} \frac{k'^2 x P(k')(1 + k'xy) - k'^3 P(k')x^2 y}{(1 + k'xy)^2}$$

$$= \frac{1}{\langle k \rangle} \sum_{k'} \frac{k'^2 x P(k')}{(1 + k'xy)^2} > 0,$$

and $\frac{\partial^2}{\partial y^2} H(x, y) = -\frac{2}{\langle k \rangle} \sum_{k'} \frac{k'^2 x P(k')}{(1+k'xy)^3} < 0$. Therefore, $H(x, y)$ is concave and monotonically increasing over y. Note that $H(x, y)$ is strictly positive. For $y_1 < y_2 \in \mathcal{Y}$ and $0 < t < 1$, we conclude that $y_1 < y_1 + t(y_2 - y_1) < y_2$. By concavity of $H(x, y)$, we obtain $H(x, y_1 + t(y_2 - y_1)) = H(x, (1 - t)y_1 + ty_2) \geq (1 - t)H(x, y_1) + tH(x, y_2)$. Dividing both sides by t yields $\frac{H(x,y_1+t(y_2-y_1))}{t} \geq \frac{(1-t)H(x,y_1)}{t} + H(x, y_2)$ which is equivalent to $\frac{H(x,y_1+t(y_2-y_1))-H(x,y_1)}{t} + H(x, y_1) \geq H(x, y_2)$. Taking the limit as $t \to 0$, we obtain

$$\frac{\partial}{\partial y_1} H(x, y_1) \geq \frac{H(x, y_2) - H(x, y_1)}{y_2 - y_1} > 0. \tag{4.32}$$

To show that there exists a constant $0 < c < 1$, such that $\frac{|H(x,y_1)-H(x,y_2)|}{|y_1-y_2|} \leq c$, $y_1, y_2 \in \mathcal{Y}$, based on (4.32), it is sufficient to show that there exists a constant $0 < c' < 1$, such that $\frac{\partial}{\partial y} H(x, y) \leq c'$, $\forall y \in \mathcal{Y}$. We have

$$\frac{\partial}{\partial y} H(x, y) = \frac{1}{\langle k \rangle} \sum_{k'} \frac{k'^2 x P(k')}{(1 + k'xy)^2} = \frac{1}{\langle k \rangle} \sum_{k'} k' P(k') \frac{k'x}{(1 + k'xy)^2},$$

and $\langle k \rangle = \sum_{k'} k' P(k')$. In addition, $\frac{k'x}{(1+k'xy)^2} = \frac{1}{\frac{1}{k'x}+2+k'xy^2} < \frac{1}{2}$, and we can conclude that $\frac{\partial}{\partial y} H(x, y) < \frac{1}{2}$.

Therefore, $|H(x, y_2) - H(x, y_1)| \leq \frac{1}{2}|y_2 - y_1|$, and $H(x, y)$ is a contraction in y uniformly over x. $\qquad\square$

Returning to the proof of Theorem 4.7, we first define $H(\psi_1, \Theta_1^*) := \frac{1}{\langle k \rangle} \sum_{k'}$ $\frac{k'^2 P(k')\psi_1 \Theta_1^*}{1+\psi_1 k' \Theta_1^*}$. From Lemma 4.1, $H(\psi_1, \Theta_1^*)$ is a contraction in Θ_1^* uniformly over

ψ_1. Therefore, for feasible s of ψ_1 and feasible $\Theta_{1,m}^*$ and $\Theta_{1,t}^*$ of Θ_1^*, we have $|H(s, \Theta_{1,m}^*) - H(s, \Theta_{1,t}^*)| \leq \frac{1}{2}|\Theta_{1,m}^* - \Theta_{1,t}^*|$. Next, let $\epsilon > 0$. Note that H is continuous in the first variable ψ_1. By choosing an appropriate $\delta > 0$ so that if $|s - t| < \delta$, where t is a feasible ψ_1, then we can have

$$|H(s, \Theta_{1,t}^*) - H(t, \Theta_{1,t}^*)| < \epsilon. \tag{4.33}$$

From Theorem 4.6, we know that $H(t, \Theta_{1,t}^*) = \Theta_{1,t}^*$ has a unique fixed-point solution. Thus, inequality (4.33) indicates that the contraction with parameter value s moves $\Theta_{1,t}^*$ a distance at most ϵ, i.e., $|H(s, \Theta_{1,t}^*) - \Theta_{1,t}^*| < \epsilon$.

Next, define a ball $\bar{B}$ with center $\Theta_{1,t}^*$ and radius r such that $|H(s, \Theta_{1,t}^*) - \Theta_{1,t}^*| \leq (1 - c)r$, where $c = \frac{1}{2}$ is the contraction constant. For $\forall w \in \bar{B}$,

$$\begin{aligned} |H(s, w) - \Theta_{1,t}^*| &\leq |H(s, w) - H(s, \Theta_{1,t}^*)| + |H(s, \Theta_{1,t}^*) - \Theta_{1,t}^*| \\ &\leq c \cdot |\Theta_{1,t}^* - w| + (1 - c) \cdot r \\ &\leq c \cdot r + (1 - c) \cdot r = r. \end{aligned}$$

Therefore, the unique solution to the fixed-point equation $\Theta_{1,s}^* = H(s, \Theta_{1,s}^*)$ lies in the ball $\bar{B}$. Since $|H(s, \Theta_{1,t}^*) - \Theta_{1,t}^*| < \epsilon$, then one choice for r is $r = \epsilon/(1 - c) = 2\epsilon$. Thus, we obtain $|\Theta_{1,s}^* - \Theta_{1,t}^*| \leq 2\epsilon$. To sum up, $|s - t| < \delta$ implies that $|\Theta_{1,s}^* - \Theta_{1,t}^*| \leq 2\epsilon$.

Hence, the Θ_1^* is continuous over ψ_1. Since $\psi_1 = \frac{\zeta_1}{(\gamma_1 + u_1)}$, then Θ_1^* is an implicit continuous function of u_1. Based on (4.29) and $\bar{I}_1^*(u_1) = \sum_k P(k)I_{1,k}^*(u_1)$, we conclude that $\bar{I}_1^*(u_1)$ is a continuous function of u_1. $\square$

Remark 4.5 Based on Theorem 4.7, the continuous mapping $\bar{I}_1^*(u_1)$ leads to a *robust* epidemic control scheme. Specifically, with a small perturbation of the unit control cost, the severity of epidemics under the optimal control resulting from (OP5) does not encounter a significant deviation.

To obtain the solution Θ_1^* with respect to ψ_1, we first denote the right hand side of (4.31) as a function of Θ_1^*, i.e., $Q : [0, 1] \to \mathbb{R}_+$. Specifically,

$$Q(\Theta_1^*) = \frac{1}{\langle k \rangle} \sum_{k'} \frac{k'^2 P(k')\psi_1 \Theta_1^*}{1 + \psi_1 k' \Theta_1^*}. \tag{4.34}$$

Then, (4.31) can be solved by using the following fixed-point iterative scheme:

$$\Theta_1^{*(n+1)} = Q(\Theta_1^{*(n)}), \quad n = 0, 1, 2, \ldots, N, \tag{4.35}$$

until $|Q(\Theta_1^{*(n+1)}) - \Theta_1^{*(n+1)}| \leq \epsilon_1$, where $\epsilon_1 > 0$ is the predefined error tolerance. Note that, from Lemma 4.1, $Q(\Theta_1^*)$ is a contraction mapping which leads to the stability and convergence of the fixed-point iterative scheme. The algorithm to obtain solution Θ_1^* is summarized in Algorithm 4.1.

For a given Θ_1^*, we have $\bar{I}_1(u_1) = \sum_k P(k) I_{1,k}(u_1)$, where $I_{1,k}(u_1) = \frac{\zeta_1 k \Theta_1^*}{\gamma_1 + u_1 + \zeta_1 k \Theta_1^*}$. Define a function $f : \mathbb{R}_+^2 \to \mathbb{R}_+$ by

$$f(\mathbf{u}) := c_1(\mathbf{u}) + c_2\big(w_1 \bar{I}_1(u_1)\big). \tag{4.36}$$

Since $\mathbf{u} \geq 0$, $c_2\big(w_1 \bar{I}_1(u_1)\big)$ is continuously differentiable, and so does $f(\mathbf{u})$. To minimize $f(\mathbf{u})$, we use the gradient descent method incorporating with backtracking line search to obtain the optimal control $\mathbf{u}^*$.

For clarity, the complete proposed method is summarized in Algorithm 4.2.

Algorithm 4.1 Fixed-Point Iterative Scheme

1: Initialize $\Theta_1^{*(0)}$, ϵ_1, $n = 0$
2: Calculate $Q(\Theta_1^{*(n)})$
3: **while** $|Q(\Theta_1^{*(n)}) - \Theta_1^{*(n)}| > \epsilon_1$ **do**
4: $\quad \Theta_1^{*(n+1)} = Q(\Theta_1^{*(n)})$
5: $\quad n = n + 1$
6: **end while**
7: **return** $\Theta_1^{*(n)}$

Algorithm 4.2 Gradient Descent Method based on Fixed-Point Iterative Scheme

---

1: Initialize the starting point $u^{(0)} = 0$, $n = 0$, tolerance ϵ_2, $u^{(-1)} = \epsilon_2 + 1$. Obtain a feasible set $\mathcal{U}$ of effort $\mathbf{u}$ from (4.24) and (4.25)
2: **while** $||\mathbf{u}^{(n)} - \mathbf{u}^{(n-1)}||_2 > \epsilon_2$ **do**
3: $\quad \psi_1^{(n)} = \frac{\zeta_1}{\gamma_1 + u_1^{(n)}}$
4: $\quad$ Obtain value $\Theta_1^{*(n)}$ through Algorithm 4.1
5: $\quad$ **for** $k = 0 : K$ **do**
6: $\qquad I_{1,k}(u_1) = \frac{\zeta_1 k \Theta_1^{*(n)}}{\gamma_1 + u_1 + \zeta_1 k \Theta_1^{*(n)}}$
7: $\quad$ **end for**
8: $\quad \bar{I}_1(u_1) = \sum_k P(k) I_{1,k}(u_1).$
9: $\quad$ Obtain $\mathbf{u}^* = \arg \min_{\mathbf{u}} c_1(\mathbf{u}) + c_2\big(w_1 \bar{I}_1(u_1)\big)$ using gradient descent method
10: $\quad \mathbf{u}_f^* = \mathrm{Proj}_{\mathcal{U}}(\mathbf{u}^*)$
11: $\quad n = n + 1$
12: $\quad \mathbf{u}^{(n)} = \mathbf{u}_f^*$
13: **end while**
14: **return** $\mathbf{u}_f^*$

Stable Exclusive Equilibrium of Strain 2
Since $\bar{I}_{1,k}^* = 0$, the optimization problem (OP3) becomes

$$(\text{OP6}): \quad \min_{\mathbf{u}} \quad c_1(\mathbf{u}) + c_2\left(w_2\bar{I}_2^*(u_2)\right)$$

$$\text{s.t.} \quad I_{2,k}^*(u_2) = \frac{\psi_2 k \Theta_2^*}{1 + \psi_2 k \Theta_2^*}, \quad \forall k \in \mathcal{K},$$

$$\psi_2 = \zeta_2/(\gamma_2 + u_2),$$

$$\text{inequalities (4.24) and (4.25)},$$

where Θ_2^* and $\bar{I}_2^*(u_2)$ are presented in (OP3). Since (OP6) is similar to (OP5), the analysis to obtain the optimal control $\mathbf{u}^*$ also follows and is omitted here.

We next comment on one observation of the optimal control effort with respect to the network structure. Different from the distribution independent strategy in disease-free regime where $\langle k \rangle$ and $\langle k^2 \rangle$ are sufficient statistics, the node degree distribution $P(k)$ plays an essential role in the optimal control of epidemics in the exclusive equilibria of strain 1 and strain 2. We summarize this result in the following corollary.

Corollary 4.4 *In the exclusive equilibria of strain 1 and strain 2 regime, the optimal control effort is distribution dependent, i.e., correlated with the node degree distribution $P(k)$, $\forall k \in \mathcal{K}$, as the epidemic severity cost c_2 depends on the average epidemic level including all nodes' degree classes.*

Remark 4.6 We have characterized the best quarantining strategy in each equilibrium regime. The next critical problem is to characterize the global optimal strategy across three equilibria. This goal can be achieved as follows. After obtaining each optimal quarantining strategy, we determine the global optimal one by comparing the objective values of three equilibria which is the one associated with the lowest cost functions of (OP4), (OP5), and (OP6).

Note that if the system operator has a predefined goal of the steady state of the network, then it is sufficient to solve one of the problems (OP4), (OP5), and (OP6). In such scenarios, the designed control is *regime-aware* by taking the control bounds in Sect. 4.2.3.1 into account.

4.2.4 Equilibria Switching via Optimal Quarantine

In this section, we present a switching phenomenon of network equilibria. Specifically, when the equilibrium state of the epidemic network without control effort is not disease-free, then it can switch to different equilibrium states through the applied control effort. To better illustrate this phenomenon, we focus on a class of symmetric control schemes and the system operator aims to suppress two epidemics jointly. Furthermore, we consider the nontrivial case $\frac{\zeta_1}{\gamma_1} \neq \frac{\zeta_2}{\gamma_2}$ where two strains of epidemics are distinguishable.

4.2.4.1 Motivation of Equilibria Switching

Before presenting the formal results, we provide an intuitive example to motivate this switching phenomenon. Recall that the optimal effort depends on the trade-off between the epidemic severity cost and the control cost captured by c_1 and c_2, respectively. Then, the steady state of epidemic network can switch if the unit cost of control effort changes. For example, the control cost of strain 1 is relatively high at the beginning which prohibits the system operator in adopting u_1 and thus the anticipated network equilibrium only contains strain 1. However, the control cost of strain 1 may decrease significantly due to the maturity of curing technology for agents infected by strain 1, and thus control effort u_1 can be applied to suppress the epidemic spreading before its outbreak. The increase of u_1 may lead to an equilibrium switching from E_2 to E_3 as the total cost of network with steady state E_3 is lower than the one stabilized at E_2, and hence it is an optimal strategy for the system operator.

4.2.4.2 Symmetric Control Effort Scenario

In general, u_1 and u_2 can admit different values. For ease of presenting the structural results, we focus on the symmetric control scenario $u_1 = u_2 = u$ and comment on the general case later in this section. This scenario is practical as the global system operator aims to suppress the spreading of two strains simultaneously. In addition, the unit cost of control effort of two strains decreases, and thus the optimal effort u increases continuously based on the continuity result in Theorem 4.7. Depending on the parameters of the epidemics, the increasing optimal control can lead to either *single* or *double* switching between equilibrium points. Based on Theorems 4.3, 4.4, and 4.5, we obtain the following corollary which presents the conditions under which the network encounters a single switching of equilibria.

Corollary 4.5 *Consider the case that $\frac{\zeta_i(k^2)}{\gamma_i(k)} > 1$, and $\frac{\zeta_i}{\gamma_i} > \frac{\zeta_{-i}}{\gamma_{-i}}$, where $i = 1$ or 2, and $-i := \{1, 2\}\backslash\{i\}$, i.e., the epidemic network is stabilized at the exclusive equilibrium of strain i without control. If*

$$\zeta_i \geq \zeta_{-i} \quad or$$

$$\zeta_i < \zeta_{-i} \ \text{ and } \ \zeta_i - \gamma_i > \zeta_{-i} - \gamma_{-i},$$

then, there exists a single transition from the exclusive equilibrium of strain i to the disease-free equilibrium with the increase of optimal u.

The single switching phenomenon in Corollary 4.5 enhances the *prediction* of network equilibrium under control, since it confirms that the exclusive equilibrium of strain $-i$ is not possible under the symmetric optimal control case in this parameter regime.

Similarly, the phenomenon of double switching of equilibrium points is presented as follows.

Corollary 4.6 *Consider the case that $\frac{\zeta_i \langle k^2 \rangle}{\gamma_i \langle k \rangle} > 1$, $i = 1, 2$, i.e., the epidemic network does not reach the disease-free equilibrium without control. When*

$$\frac{\zeta_i}{\gamma_i} > \frac{\zeta_{-i}}{\gamma_{-i}}, \quad \zeta_i < \zeta_{-i}, \quad \text{and} \quad \frac{\zeta_i - \gamma_i}{\zeta_{-i} - \gamma_{-i}} < 1,$$

where $i = 1, 2$ and $-i := \{1, 2\}\backslash\{i\}$, then, there exist transitions from the exclusive equilibrium of strain i, to the exclusive equilibrium of strain $-i$, and to the disease-free equilibrium with the increase of u.

For the special case that $\frac{\zeta_1}{\gamma_1} = \frac{\zeta_2}{\gamma_2}$, and $\frac{\zeta_i \langle k^2 \rangle}{\gamma_i \langle k \rangle} > 1$, $i = 1, 2$, when $\zeta_i > \zeta_{-i}$, there exist transitions from the current network equilibrium (mixed steady state with both strains) to the exclusive equilibrium of strain i, and then to the disease-free equilibrium with the increase of optimal control u as the unit control cost decreases.

To identify the optimal policies under which the control effort leads to a stable disease-free equilibrium through switching, we present the following definition.

Definition 4.1 (*Fulfilling Threshold*) The fulfilling threshold refers to the optimal control $\bar{\mathbf{u}}^o = (\bar{u}_1, \bar{u}_2)$ under which the epidemic network stabilizes at the disease-free equilibrium after switching of network equilibria, and the total cost $c_1(\bar{\mathbf{u}})$ is the lowest among all control policies. Equivalently, $\bar{\mathbf{u}}$ satisfies the following conditions:

$$c_1(\bar{\mathbf{u}}) \leq c_1(\mathbf{u}), \quad \forall \mathbf{u},$$

$$\bar{u}_1 \geq \frac{\zeta_1 \langle k^2 \rangle}{\langle k \rangle} - \gamma_1, \tag{4.37}$$

$$\bar{u}_2 \geq \frac{\zeta_2 \langle k^2 \rangle}{\langle k \rangle} - \gamma_2.$$

Based on Definition 4.1, we next characterize the fulfilling threshold in the investigated scenario.

Proposition 4.1 *The optimal control effort does not increase after the epidemic network switches from the exclusive equilibrium E_2 or E_3 to the disease-free equilibrium E_1. In the investigated symmetric control scenario with constraint $u_1 = u_2$, the fulfilling threshold is*

$$\bar{u} = \max\left(0, \frac{\zeta_1 \langle k^2 \rangle}{\langle k \rangle} - \gamma_1, \frac{\zeta_2 \langle k^2 \rangle}{\langle k \rangle} - \gamma_2\right). \tag{4.38}$$

Proof The fulfilling threshold in the studied cases can be directly verified by the zero epidemic cost in regime E_1 and the monotonically increasing function c_1 with respect to the applied effort. Based on Definition 4.1 and symmetric control structure, we can obtain the threshold $\bar{u}$ in (4.38). $\qquad\square$

Remark 4.7 The fulfilling threshold in Proposition 4.1 provides an upper bound for the network operator's control effort to bring the network equilibria to the disease-free regime. As the unit cost of effort decreases, the amount of optimal control should not exceed the fulfilling threshold.

Another result on the number of network equilibria switching is summarized as follows.

Corollary 4.7 *Under the symmetric optimal control scenario with decreasing unit control cost, the maximum number of network equilibria switching is two.*

Corollary 4.7 generalizes Corollaries 4.5 and 4.6 by studying the entire parameter regime. The monotonically increasing optimal control yields either single or double switching of equilibria. For general cases in which optimal u_1 and u_2 are not necessarily the same, then the switching of network equilibria depends on the specific unit costs of u_1 and u_2. However, if the system operator has a preference to avoid the outbreak of strain i, then as the optimal control u_i increases, either single or double switching happens with the network stabilizing at disease-free equilibrium depending on the epidemic system parameters.

4.2.5   Case Studies

In this section, we corroborate the obtained results with numerical experiments. First, we generate a scale-free network with 500 nodes using the Barabási–Albert model [38]. The degree distribution of the network satisfies $P(k) \sim k^{-3}$. The typical generated random network in the following studies has an average connectivity $\langle k \rangle = 1.996$ and $\langle k^2 \rangle = 13.75$. Our objective is to design the optimal control of interdependent epidemics spreading under different network equilibrium cases.

The functions in the optimization problems admit the forms: $c_1(\mathbf{u}) = K_1 u_1 + K_2 u_2$, and $c_2(w_1 \bar{I}_1^*(u) + w_2 \bar{I}_2^*(u)) = K_3(\bar{I}_1^*(u) + \bar{I}_2^*(u))$, where K_1, K_2 and K_3 are positive constants, and $w_1 = w_2 = 1$. Specifically, we choose $K_1 = 15$, $K_2 = 10$ and $K_3 = 50$. For better illustration purposes, we assume that strain 1 and strain 2 have the same spreading rate, i.e., $\zeta_1 = \zeta_2 = \zeta$. We find and compare the optimal control solutions of the following two scenarios: scenario I where $\gamma_1 = 0.5$, $\gamma_2 = 0.3$, and scenario II where $\gamma_1 = 0.5$, $\gamma_2 = 0.8$.

4.2.5.1   Optimal Control in Disease-Free Case

In the disease-free case, the epidemic spreading levels are zero at the steady state. By solving (OP4), we obtain the results of optimal control which are shown in Fig. 4.1. We can see that the control efforts u_1 and u_2 both increase linearly with the spreading rate ζ as expected by (4.28). Due to the same recovery rates of strain 1 in two scenarios, the applied control efforts u_1 overlap as shown in Fig. 4.1a. In addition,

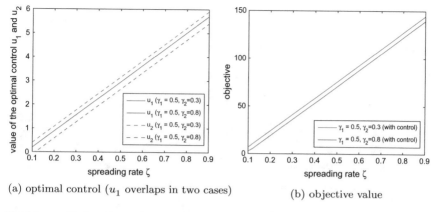

(a) optimal control (u_1 overlaps in two cases) (b) objective value

Fig. 4.1 **a** and **b** show the results of the optimal control and the associated objective value, respectively, where the network stabilizes at the disease-free equilibrium

because of a smaller self-recovery rate of strain 2 in scenario I, its corresponding control effort u_2 is larger than that in scenario II. Hence, the optimal objective value in scenario II is smaller than that of scenario I shown in Fig. 4.1b.

4.2.5.2 Optimal Control in Exclusive Equilibrium Case

We investigate the case when the network is stabilized at the exclusive equilibrium of strain 1. By solving (OP5) using the proposed Algorithm 4.2, the obtained results are shown in Fig. 4.2. Specifically, Fig. 4.2a, b show the optimal control efforts. In scenario I, the control u_1 (red line in Fig. 4.2a) increases first when the spreading rate ζ is relatively small. It then decreases after $\zeta > 0.55$, since it is not economical to control the spreading of strain 1 comparing with its control cost. Further, because the recovery rate of strain 2 in scenario I is low, the applied control u_2 (red dotted line in Fig. 4.2b) should be relatively large to suppress its spreading. An important phenomenon is that u_2 decreases after $\zeta > 0.55$, which follows the pattern of u_1, since u_2 can be chosen as long as it satisfies the conditions in Theorem 4.4, and strain 2 does not exist at the steady state. In scenario II, due to the high self-recovery rate γ_2, strain 2 dies out at the equilibrium even without control. Thus, the control of strain 2 is 0, i.e., $u_2 = 0$ (blue dotted line in Fig. 4.2a). In addition, the control u_1 in this scenario (blue line in Fig. 4.2a) first increases to compensate the spreading of strain 1. Then, it stays flat after $\zeta > 0.27$, since otherwise larger control u_1 leads to a network equilibrium switching from E_2 to E_3. Figure 4.2c depicts the severity of epidemics at the steady state with and without control. We can conclude that the optimal control effectively reduces the spreading of epidemics in both scenarios. Note that the epidemic spreading levels without the control intervention overlap in two cases (dotted lines in Fig. 4.2c) though only strain 2 and strain 1 exist at equilibrium in scenarios I and II, respectively. The reason is that the severity of epidemics is

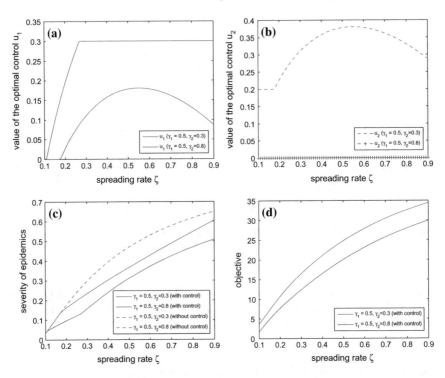

Fig. 4.2 The network is stabilized at the exclusive equilibrium of strain 1. **a** and **b** are the optimal control of strain 1 and strain 2, respectively. **c** and **d** show the severity of epidemics and the corresponding objective value under the optimal control, respectively

determined by the network structure and the steady state, while the parameter ζ only influences the rate of epidemics spreading.

4.2.5.3   Transition of the Equilibrium Through Control

In this section, we illustrate the transition between the epidemic equilibrium through control. First, we study the single transition case. From Corollary 4.5, we choose $\zeta_1 = 0.2$, $\gamma_1 = 0.4$, $\zeta_2 = 0.15$ and $\gamma_2 = 0.4$. The result is shown in Fig. 4.3. As the unit control cost changes, the network equilibrium at steady state will be different. Specifically, as the optimal control increases due to the decrease of unit control cost, the epidemic network equilibrium switches from the exclusive equilibrium of strain 2 to the disease-free equilibrium. For the double transitions case, based on Corollary 4.6, we select parameters $\zeta_1 = 0.1$, $\gamma_1 = 0.1$, $\zeta_2 = 0.15$ and $\gamma_2 = 0.2$. The result is shown in Fig. 4.4. Consistent with Corollary 4.6, the network equilibrium switches first from the exclusive equilibrium of strain 2 to the exclusive equilibrium of strain 1, and then to the disease-free equilibrium, as the applied optimal control

Fig. 4.3 Transition of the equilibrium with the increase of control across two regimes: from the exclusive equilibrium of strain 2 (III) to the disease-free equilibrium (I)

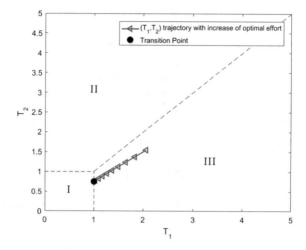

Fig. 4.4 Transition of the equilibrium with the increase of control across three regimes. (from the exclusive equilibrium of strain 2 (III) to the exclusive equilibrium of strain 1 (II), then to the disease-free equilibrium (I))

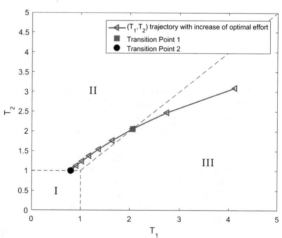

increases. One common feature in these two cases is that once the effort drives the network to the disease-free equilibrium, the control effort ceases to increase, where fulfilling threshold is reached (corresponding to the effort level at the transition point denoted by black dot in Figs. 4.3 and 4.4). Specifically, based on Proposition 4.1, the fulfilling thresholds in Figs. 4.3 and 4.4 are 0.978 and 0.834, respectively.

4.3 Summary and Notes

In this chapter, we have studied the optimal control of interdependent epidemics spreading over complex networks. The competing mechanism between two strains of epidemics results in a non-coexistence phenomenon at the steady state. Furthermore, we have explicitly derived the conditions under which the network is stabilized at different equilibria with control. The optimal control computed via the designed iterative algorithm can effectively reduce the spreading of epidemics. At the disease-free equilibrium, the optimal control is independent of nodes' degree distribution as the optimal strategy can be fully determined by the sufficient statistics including the average degree and the second moment of the degree distribution. Furthermore, depending on the epidemic parameters, the network equilibrium can switch via the adopted control strategy. Once the epidemic network switches to the disease-free equilibrium under the optimal control, the applied effort does not increase though the unit cost of effort continues to decrease, and the optimal control effort at the associated switching point is called the fulfilling threshold.

The readers interested in an overview of epidemic processes in complex networks can refer to [2] for more details. In addition, the researchers interested in the control of epidemics can refer to [19, 22, 27–30].

References

1. Pastor-Satorras R, Vespignani A (2001) Epidemic spreading in scale-free networks. Phys Rev Lett 86(14):3200
2. Pastor-Satorras R, Castellano C, Van Mieghem P, Vespignani A (2015) Epidemic processes in complex networks. Rev Mod Phys 87(3):925–979
3. Chen J, Touati C, Zhu Q (2019) Optimal secure two-layer iot network design. IEEE Trans Control Netw Syst. https://doi.org/10.1109/TCNS.2019.2906893
4. Chen J, Zhu Q (2016) Resilient and decentralized control of multi-level cooperative mobile networks to maintain connectivity under adversarial environment. In: Conference on decision and control (CDC). IEEE, pp 5183–5188
5. Chen J, Zhu Q (2016) Interdependent network formation games with an application to critical infrastructures. In: American control conference (ACC). IEEE, pp 2870–2875
6. Moreno Y, Nekovee M, Pacheco AF (2004) Dynamics of rumor spreading in complex networks. Phys Rev E 69(6):066,130
7. Omic J, Orda A, Mieghem PV (2009) Protecting against network infections: a game theoretic perspective. In: IEEE conference on computer and communications, pp 1485–1493
8. Garetto M, Gong W, Towsley D (2003) Modeling malware spreading dynamics. IEEE Conf Comput Commun 3:1869–1879
9. Gross T, DLima CJD, Blasius B (2006) Epidemic dynamics on an adaptive network. Phys Rev Lett 96(20):208,701
10. Pastor-Satorras R, Vespignani A (2002) Immunization of complex networks. Phys Rev E 65(3):036,104
11. Chen J, Zhang R, Zhu Q (2017) Optimal control of interdependent epidemics in complex networks. In: SIAM workshop on network science
12. Van Mieghem P, Omic J, Kooij R (2009) Virus spread in networks. IEEE/ACM Trans Netw 17(1):1–14
13. Gang Y, Tao Z, Jie W, Zhong-Qian F, Bing-Hong W (2005) Epidemic spread in weighted scale-free networks. Chin Phys Lett 22(2):510
14. Newman ME (2002) Spread of epidemic disease on networks. Phys Rev E 66(1):016,128
15. Dickison M, Havlin S, Stanley HE (2012) Epidemics on interconnected networks. Phys Rev E 85(6):066,109
16. Saumell-Mendiola A, Serrano MÁ, Boguná M (2012) Epidemic spreading on interconnected networks. Phys Rev E 86(2):026,106

17. Paré PE, Beck CL, Nedić A (2018) Epidemic processes over time-varying networks. IEEE Trans Control Netw Syst 5(3):1322–1334
18. Prakash BA, Tong H, Valler N, Faloutsos M, Faloutsos C (2010) Virus propagation on time-varying networks: theory and immunization algorithms. In: Joint European conference on machine learning and knowledge discovery in databases. Springer, pp 99–114
19. Hansen E, Day T (2011) Optimal control of epidemics with limited resources. J Math Biol 62(3):423–451
20. Sahneh FD, Scoglio C (2011) Epidemic spread in human networks. In: IEEE conference on decision and control and European control conference (CDC-ECC), pp 3008–3013
21. Ramirez-Llanos E, Martinez S (2014) A distributed algorithm for virus spread minimization. In: American control conference (ACC), pp 184–189
22. Befekadu GK, Zhu Q (2019) Optimal control of diffusion processes pertaining to an opioid epidemic dynamical model with random perturbations. J Math Biol 78(5):1425–1438
23. Trajanovski S, Hayel Y, Altman E, Wang H, Van Mieghem P (2015) Decentralized protection strategies against sis epidemics in networks. IEEE Trans Control Netw Syst 2(4):406–419
24. Hayel Y, Zhu Q (2017) Epidemic protection over heterogeneous networks using evolutionary poisson games. IEEE Trans Inf Forensics Secur 12(8):1786–1800
25. Venturino E (2001) The effects of diseases on competing species. Math Biosci 174(2):111–131
26. Han L, Pugliese A (2009) Epidemics in two competing species. Nonlinear Anal Real World Appl 10(2):723–744
27. Karrer B, Newman M (2011) Competing epidemics on complex networks. Phys Rev E 84(3):036,106
28. Taynitskiy V, Gubar E, Zhu Q (2017) Optimal impulse control of bi-virus SIR epidemics with application to heterogeneous internet of things. In: IEEE constructive nonsmooth analysis and related topics (CNSA), pp 1–4
29. Lee P, Clark A, Alomair B, Bushnell L, Poovendran R (2018) Adaptive mitigation of multi-virus propagation: a passivity-based approach. IEEE Trans Control Netw Syst 5(1):583–596
30. Gubar E, Zhu Q (2013) Optimal control of influenza epidemic model with virus mutations. In: European control conference (ECC), IEEE, pp 3125–3130
31. Chen J, Zhu Q (2019) Interdependent strategic security risk management with bounded rationality in the internet of things. IEEE Trans Inf Forensics Secur. https://doi.org/10.1109/TIFS.2019.2911112
32. Chen J, Zhu Q (2018) A linear quadratic differential game approach to dynamic contract design for systemic cyber risk management under asymmetric information. In: 2018 56th annual Allerton conference on communication, control, and computing (Allerton). IEEE, pp 575–582
33. Pawlick J, Zhu Q (2017) Strategic trust in cloud-enabled cyber-physical systems with an application to glucose control. IEEE Transa Inf Forensics Secur 12(12):2906–2919
34. Pawlick J, Chen J, Zhu Q (2019) iSTRICT: An interdependent strategic trust mechanism for the cloud-enabled internet of controlled things. IEEE Trans Inf Forensics Secur 14(6):1654–1669
35. Chen J, Zhu Q (2016) Optimal contract design under asymmetric information for cloud-enabled internet of controlled things. In: International conference on decision and game theory for security. Springer, pp 329–348
36. Chen J, Zhu Q (2017) Security as a service for cloud-enabled internet of controlled things under advanced persistent threats: a contract design approach. IEEE Trans Inf Forensics Secur 12(11):2736–2750
37. Chen J, Touati C, Zhu Q (2017) A dynamic game analysis and design of infrastructure network protection and recovery. ACM SIGMETRICS Perform Eval Rev 45(2):125–128
38. Barabási AL, Albert R (1999) Emergence of scaling in random networks. Science 286(5439):509–512

Chapter 5
Optimal Secure Interdependent Infrastructure Network Design

5.1 Interdependent Infrastructure Network Security

In this chapter, we adopt the established model of network-of-networks to design optimal secure interdependent infrastructures. IoTs have witnessed a tremendous development with a variety of applications, such as virtual reality, intelligent supply chain, and smart home. In this highly connected world, IoT devices are massively deployed and connected to cellular or cloud networks. For example, in smart grids, wireless sensors are adopted to collect the data of buses and power transmission lines [1]. The collected data can then be sent to a supervisory control and data acquisition (SCADA) center through cellular networks for grid monitoring and decision planning purposes [2, 3]. Smart home is another example of IoT application. Various devices and appliances in a smart home including air conditioner, lights, TV, tablets, refrigerator and smart meter are interconnected through the cloud, improving the quality of the living.

IoT networks can be viewed as multi-layer networks with the existing infrastructure networks (e.g., cloud and cellular networks) and the underlaid device networks. The connectivity of IoT networks plays an important role in information dissemination. On the one hand, devices can communicate directly with other devices in the underlaid network for local information. On the other hand, devices can also communicate with the infrastructure networks to maintain a global situational awareness. In addition, for IoT devices with insufficient on-board computational resources such as wearables and drones, they can outsource heavy computations to the data centers through cloud networks, and hence extend the battery lifetime [4]. Vehicular network is an illustrative example for understanding the two-tier feature of IoT networks [5]. In an intelligent transportation network, vehicle-to-vehicle (V2V) communications enable two vehicles to communicate and exchange information, e.g., accidents, speed alerts, notifications. In addition, vehicles can also communicate with roadside infrastructures or units (RSU) that belong to one or several service providers for exchanging various types of data related to different applications including GPS navigation,

© The Author(s), under exclusive license to Springer Nature Switzerland AG 2020
J. Chen and Q. Zhu, *A Game- and Decision-Theoretic Approach to Resilient Interdependent Network Analysis and Design*, SpringerBriefs in Control, Automation and Robotics, https://doi.org/10.1007/978-3-030-23444-7_5

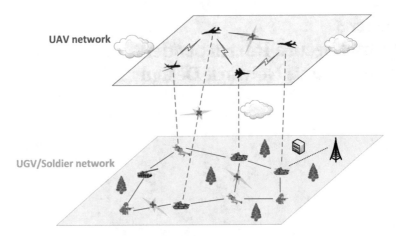

Fig. 5.1 In IoBT networks, a team of UAVs and a group of soldiers and UGVs execute missions cooperatively. The agents in the battlefield share critical information through D2D communications. The UAV network and ground network form a two-layer network which faces cyber threats, e.g., jamming attacks which can lead to link removals

parking and highway tolls inquiry. In this case, the vehicles form one network while the infrastructure nodes form another network. Due to the interconnections between two networks, vehicles can share information through infrastructure nodes or by direct V2V communications.

IoT communication networks are vulnerable to cyber attacks including the denial-of-service (DoS) and jamming attacks [6, 7]. To compromise the communication between two specific devices, the attacker can adopt the selective jamming attack [8]. More specifically, the attacker selectively targets specific channels and packets which disrupts the communications by transmitting a high-range or high-power interference signal. This adversarial behavior leads to communication link removals in IoT network. Therefore, to maintain the connectivity of devices, IoT networks need to be secure and resistant to malicious attacks. For example, V2V communication links of a car can be jammed, and hence the car loses the real-time traffic information of the road which may further cause traffic delays and accidents especially in the futuristic self-driving applications. Hence, IoT networks should be constructed in a tactic way by anticipating the cyber attacks. Internet of Battlefield Things (IoBT) is another example of mission-critical IoT systems [9]. As depicted in Fig. 5.1, in IoBT networks, a team of unmanned aerial vehicles (UAVs) serves as one layer of wireless relay nodes for a team of unmanned ground vehicles (UGVs) and soldiers equipped with wearable devices to communicate between themselves or exchange critical information with the command-and-control nodes. The UAV network and the ground network naturally form a two-layer network in a battlefield which can be susceptible to jamming attacks. It is essential to design communication networks that can allow the IoBT networks to be robust to natural failures and secure to cyber attacks in order to keep a high-level situational awareness of agents in a battlefield.

Due to heterogeneous and multi-tier features of the IoT networks, the required security levels can vary for different networks. For example, in IoBT networks, the connectivity of UAV networks requires a higher security level than the ground network if the UAVs are more likely to be targeted by the adversary. Similarly, in vehicular networks, the communication links between RSUs need a high-level protection when they anticipate more attacks than the vehicles do. Therefore, it is imperative to design secure IoT networks resistant to link attacks and maintain the two-layer network connectivity with heterogeneous security requirements simultaneously.

5.2   Optimal Secure Two-Layer Network Design with an Application to IoBT

We present a heterogeneous IoT network design framework in which network links are vulnerable to malicious attacks. To enhance the security and the robustness of the network, an IoT network designer can add extra links to provide additional communication paths between two nodes or secure links against failures by investing resources to protect the links. To allocate links, note that when the nodes in the IoT network are within a short distance, then the classical wireless communication technologies can be adopted including WiFi, Bluetooth, and Zigbee. In comparison, when the distance is large, then one option that has recently emerged is called ultra narrow band (UNB) [10] that uses the random frequency and time multiple access [11]. The UNB is dedicated for mission-critical IoT systems for providing reliable communication services in long range. The goal of the multi-tier network design is to make the network connectivity resistant to link removal attacks by anticipating the worst attack behaviors. Different from previous works [12, 13] which have focused on the secure design of single-layer networks, in this chapter, the network designer needs to take into account the heterogeneous features of the IoT networks by imposing different security requirements on each layer which presents a new set of challenges for network design.

We focus on a two-layer IoT network and aim to design each network resistant to different number of link failures with minimum resources [6]. We characterize the optimal strategy of the secure network design problem by first developing a lower bound on the number of links a secure network requires for a given budget of protected links. Then, we provide necessary and sufficient conditions under which the bounds are achieved and present a method to construct an optimal network that satisfies the heterogeneous network design specifications with the minimum cost. Furthermore, we characterize the robust network topologies which optimally satisfy a class of security requirements. These robust optimal networks are applicable to the cases when the cyber threats are not perfectly perceived or change dynamically, typically happening in the mission-critical scenarios when the attacker's action is partially observable.

Finally, we use IoBT as a case study to illustrate the analytical results and obtain insights in designing secure networks. We consider a mission-critical battlefield scenario in which the UAV network anticipates higher cyber threats than the soldier network, and the number of UAVs is less than the number of soldiers. We observe that as the cost of forming a protected communication link becomes smaller, more secure connections are formed in the optimal IoBT network. In addition, the designed network is resilient to the change of agents in the battlefield. For example, when a group of soldiers join in the battle, they only need to connect to a set of neighboring companions, which is convenient to implement in practice. For the scenarios in which soldiers leave the battlefield, the optimal network can be reconfigured in a similar fashion, i.e., those agents who lose some degree of communications should build up new connections with some other neighbors to stay resistant to attacks. We also study the reconfiguration and resilience of the UAV network as nodes leave and join the battlefield.

Related Work

Due to the increasing cyber threats, IoT security becomes a critical concern [14–16]. Depending on the potential of cyber attackers, IoT networks face heterogeneous types of attacks [17]. For example, attackers can target the edge computing nodes in IoT, e.g., RFID readers and sensor nodes. Some typical adversarial scenarios include the node replication attack by replicating one node's identification number [18], DoS by battery draining, sleep deprivation, and outage attacks [19, 20]. The attackers can also launch attacks through the IoT communication networks [21–23]. Quintessential examples include the eavesdropping attack where the attacker captures the private information over the channel, and utilizes the information to design other tailored attacks [24]. Another example is the data injection attack where the attacker can inject fraudulent packets into IoT communication links through insertion, manipulation, and replay techniques [25]. The work in this chapter is also related to the interdependent infrastructure networks in which their secure and resilient operational strategies are critical in ensuring the system performance [26–30].

To mitigate the cyber threats in IoT, a large number of works have focused on addressing the security issues by using different methodologies [7, 31]. The authors in [4, 32, 33] have proposed a switching control method to enhance the security and resilience of cloud-enabled systems. A contract-theoretic approach has been adopted to guarantee the performance of security services in the Internet of controlled things and networked systems [21, 34, 35]. Farooq and Zhu [9] have designed a reconfigurable communication network for information dissemination in IoBT using an epidemic model. Chen et al. [22] have proposed a dynamic game model including pre-attack defense and post-attack recovery phases in designing resilient IoT-enabled infrastructure networks. In [36], the authors have studied the resilience aspect of routing problem in parallel link communication networks using a two-player game framework and designed stable algorithms to compute the equilibrium strategy. The authors in [37] have developed an adaptive strategic cyber defense for mitigating advanced persistent threats in infrastructures.

We investigate the secure design of IoT network by considering its connectivity measure [12, 13, 38–41] through the lens of graph theory [42]. Comparing with the previous works [12, 13] that have focused on a single-layer adversarial network design, we model the IoT as a two-layer network and strategically design each layer of the network with heterogeneous security requirements.

5.2.1 Heterogeneous Two-Layer IoT Network Design Formulation

In this section, we formulate a two-layer secure IoT network design problem. Due to the heterogeneous features of IoT networks, the devices at each layer face different levels of cyber threats. To maintain the global situational awareness, the designer aims to devise an IoT network with a minimum cost, where each layer of IoT network should remain connected in the presence of a certain level of adversarial attacks.

Specifically, we model the two-layer IoT network with two sets of devices or nodes[1] denoted by S_1 and S_2. Each set of nodes is of a different type. Specifically, denote by $n_1 := |S_1|$ and $n_2 := |S_2|$ the number of nodes of type 1 and 2, respectively, where $|\cdot|$ denotes the cardinality of a set. We unify them to $n = n_1 + n_2$ vertices that are numbered from 1 to n starting from nodes in S_1. Thus, a node labeled i is of type 1 if and only if $i \leq n_1$. Note that each set of nodes forms an IoT subnetwork. Together with the interconnections between two sets of nodes, the subnetworks form a two-layer IoT network. Technically, the communication protocols between nodes within and across different layers can be either the same or heterogeneous depending on the adopted technology by considering the physical distance constraints. Furthermore, the nodes' functionality can be different in two subnetworks depending on their specific tasks. In this chapter, our focus lies in the high-level of network connectivity maintenance.

In standard graph theory, an *edge* (or a *link*) is an unordered pair of vertices: $(i, j) \in [\![1, n]\!]^2, i \neq j$, where $[\![1, n]\!]^2$ is a set including all the pairs of integers between 1 and n. We recall that two vertices (nodes) i_0 and i_L are said *connected* in a graph of nodes $S_1 \cup S_2$ and a set of edges $\mathcal{E}$ if there exists a path between them, i.e., a finite alternating sequence of nodes and distinct links: $i_0, (i_0, i_1), i_1, (i_1, i_2), i_2, \ldots, (i_{L-1}, i_L), i_L$, where $i_l \in S_1 \cup S_2$ and $(i_{l-1}, i_l) \in \mathcal{E}$ for all $1 \leq l \leq L$.

In our IoT networks, the communication links (edges) are vulnerable to malicious attacks, e.g., jamming and DoS, which result in link removals. To keep the IoT network resistant to cyber attacks, the network designer can either invest (i) in redundancy of the path, i.e., using extra links so that two nodes can communicate through different paths, or (ii) in securing its links against failures where we refer to these special communication edges as *protected links*. These protected links can be typically designed using moving target defense (MTD) strategies, where the

---

[1]Nodes and vertices in the IoT network refer to the devices, and they are used interchangeably. Similar for the terms edges and links.

designer randomizes the usage of communication links among multiple created channels between two nodes [43]. More precisely, we consider that for the designer, the cost per non-protected link created is c_{NP} and the cost per protected link created is c_P. It is natural to have $c_{NP} \leq c_P$ since creation of a protected link is more costly than that of a non-protected one. For clarity, we assume that the costs of protected or non-protected links at two different layers are the same. If the costs of creating links are different in two subnetworks, then the network designer needs to capture this link creation difference in his objective [44]. Let $\mathcal{E}_{NP} \subseteq \mathcal{E}$ be the set of non-protected links and $\mathcal{E}_P \subseteq \mathcal{E}$ be the set of protected links in the IoT network, and $\mathcal{E}_{NP} \cup \mathcal{E}_P = \mathcal{E}$. We assume that the protection is perfect, i.e., links will not fail under attacks if they are protected. Therefore, an adversary does not have an incentive to attack protected links. Denote the strategy of the attacker by $\mathcal{E}_A$, then it is sufficient to consider attacks on a set of links $\mathcal{E}_A \subseteq \mathcal{E}_{NP}$. Furthermore, we assume that the network designer can allocate links between any nodes in the network. In the scenarios that setting up communication links between some nodes is not possible, then the network designer needs to take into account this factor as constraints when designing networks.

The heterogeneous features of IoT networks naturally lead to various security requirements for devices in each subnetwork. Hence, we further consider that the nodes in IoT network have different criticality levels (k_1 and k_2 for nodes of type 1 and 2, respectively, with $k_1, k_2 \in [\![0, |\mathcal{E}_{NP}|]\!]$, where $[\![a, b]\!]$ denotes a set of integers between a and b). It means that subnetworks 1 and 2 should remain connected after the compromise of *any* k_1 and k_2 links in $\mathcal{E}_{NP}$, respectively. Thus, the designer needs to prepare for the worst case of link removal attacks when designing the two-layer IoT network. Our problem is beyond the robust network design where the link communication breakdown is generally caused by nature failures. In this chapter, we consider the link removal which is a consequence of cyber attacks, e.g., jamming and DoS attack. Furthermore, in our problem formulation, the network designer can allocate protected links which can be seen as a security practice, and he takes into account the strategic behavior of attackers, and designs the optimal secure networks. Without loss of generality, we have the following two assumptions:

(A1) $k_1 \leq k_2$.
(A2) $n_1 \geq 1, n_2 \geq 1$.

Specifically, (A1) indicates that the IoT devices in subnetwork 2 are relatively more important than those in subnetwork 1, and thus subnetwork 2 should be more resistant to cyber attacks. Another interpretation of (A1) can also be that subnetwork 2 faces a higher level of cyber threats, and the network designer needs to prepare a higher security level for subnetwork 2. In addition, (A2) ensures that no IoT subnetwork is empty.

More precisely, consider a set of vertices $\mathcal{S}_1 \cup \mathcal{S}_2$ and edges $\mathcal{E}_P \cup \mathcal{E}_{NP}$. The IoT network designer needs to guarantee the following two cases:

(a) if $|\mathcal{E}_A| \leq k_1$, then all nodes remain attainable in the presence of attacks, i.e., $\forall i, j \in \mathcal{S}_1 \cup \mathcal{S}_2$, there exists a path in the graph $(\mathcal{S}_1 \cup \mathcal{S}_2, \mathcal{E}_P \cup \mathcal{E}_{NP} \setminus \mathcal{E}_A)$ between i and j.

(b) if $|\mathcal{E}_A| \leq k_2$, nodes of type 2 remain attainable after attacks, i.e., $\forall i, j \in \mathcal{S}_2$, there exists a path in the graph $(\mathcal{S}_1 \cup \mathcal{S}_2, \mathcal{E}_P \cup \mathcal{E}_{NP} \setminus \mathcal{E}_A)$ between i and j.

Remark 5.1 We denote the designed network satisfying (a) and (b) above by $s^D := (\mathcal{S}_1 \cup \mathcal{S}_2, \mathcal{E}_P \cup \mathcal{E}_{NP})$, and call such heterogeneous IoT networks (k_1, k_2)-*resistant* (with $k_1 \leq k_2$). The proposed $(k1, k2)$-resistant metric provides a flexible network design guideline by specifying various security requirements on different network components. Furthermore, we care about each node's degree which requires an explicit agent-level quantification. Then, the (k_1, k_2)-resistant metric is more preferable than measure of the proportion of links in each subnetwork, where the latter metric only gives a macroscopic description of the link allocation over two subnetworks.

Given the system's parameters $\mathcal{S}_1$, $\mathcal{S}_2$, k_1, and k_2, an optimal strategy for the IoT network designer is the choice of a set of links $\mathcal{E}_P \cup \mathcal{E}_{NP}$ which solves the optimization problem:

$$\min_{\mathcal{E}_P, \mathcal{E}_{NP}} \quad c_P |\mathcal{E}_P| + c_{NP} |\mathcal{E}_{NP}|$$
$$\text{s.t.} \ \mathcal{E}_P \subseteq [\![1, n]\!]^2, \ \mathcal{E}_{NP} \subseteq [\![1, n]\!]^2,$$
$$\mathcal{E}_P \cap \mathcal{E}_{NP} = \emptyset,$$
$$s^D = (\mathcal{S}_1 \cup \mathcal{S}_2, \mathcal{E}_P \cup \mathcal{E}_{NP}) \text{ is } (k_1, k_2)-\text{resistant.}$$

From the above optimization problem, the optimal network design cost directly depends on c_P and c_{NP}. In addition, as we will analyze in Sect. 5.2.2, the cost ratio $\frac{c_P}{c_{NP}}$ plays a critical role in the optimal strategy design.

Under the optimal design strategy, compromising a node with low degree, i.e., k_1 degree in subnetwork 1 and k_2 degree in subnetwork 2, is not feasible for the attacker, since the degree of any nodes without protected link in the network is larger than k_1 or k_2 depending on the nodes' layers.

Note that the above designer's constrained optimization problem is not straightforward to solve. First, the size of search space increases exponentially as the number of nodes in the IoT network grows. Therefore, we need to find a scalable method to address the optimal network design. Second, the heterogeneous security requirements make the problem more difficult to solve. On the one hand, two subnetworks are separate since they have their own design standards. On the other hand, we should tackle these two layers of network design in a holistic fashion due to their natural couplings.

5.2.2 Analytical Results and Optimal IoT Network Design

In this section, we provide an analytical study of the designer's optimal strategy, i.e., the optimal two-layer IoT network design.

We first develop, for given system parameters S_1, S_2, k_1, k_2, c_P and c_{NP}, and for each possible number of protected links $p = |\mathcal{E}_P|$, a lower bound on the number of non-protected links that have any (k_1, k_2)-resistant network with p protected links (Sect. 5.2.2.1). Then, we study three important cases, namely when p takes values 0, $n_2 - 1$ and $n_1 + n_2 - 1$, and present for each of them sufficient conditions under which the lower bounds are attained (Sect. 5.2.2.2). Based on this study, we can obtain the main theoretical results of this chapter, which include the optimal strategy for the designer, i.e., a (k_1, k_2)-resistant IoT network with the minimal cost, as well as the robust optimal strategy, and constructive methods of an optimal IoT network (Sect. 5.2.2.3).

5.2.2.1 A Lower Bound on the Number of (Non-protected) Links

Recall that the system parameters are S_1, S_2, k_1, k_2, c_P and c_{NP} (corresponding to the set of nodes of criticality level 1 and 2, the values of criticality, and the unitary cost of creating protected and non-protected links). We first address the question of a lower bound on the cost for the designer with an additional constraint on the number of protected links p in the network. Since the cost is linear with the number of non-protected links, it amounts to finding a lower bound on the number of non-protected links that are required in any (k_1, k_2)-resistant network with p protected links.

Let $\tilde{s}_p^D$ be a (k_1, k_2)-resistant network containing p protected links. Then, we have the following proposition on the lower bound $|\mathcal{E}_{NP}|$.

Proposition 5.1 (Lower bound on $|\mathcal{E}_{NP}|$) *The number of non-protected links of $\tilde{s}_p^D$ is at least of*

$$(i) \quad \frac{n_1(k_1 + 1) + (n_2 - p)(k_2 + 1)}{2}, \qquad\qquad if\ 0 \le p \le n_2 - 2,$$

$$(ii) \quad \frac{(n - p)(k_1 + 1)}{2}, \qquad\qquad if\ n_2 - 1 \le p \le n_1 + n_2 - 2,$$

$$(iii)\ 0, \qquad\qquad\qquad if\ p = n_1 + n_2 - 1.$$

Note that p takes integer values in each regime. The results are further illustrated in Fig. 5.2.

Before proving Proposition 5.1, we first present the notion of network contraction in the following.

Network Contraction: Let $g = (S_1 \cup S_2, \mathcal{E}_P \cup \mathcal{E}_{NP})$ be a network. Given a link $(i, j) \in \mathcal{E}_P$, the network denoted by $g \oslash (i, j)$ refers to the one obtained by contracting the link (i, j); i.e., by merging the two nodes i and j into a single node $\{i, j\}$ (supernode). Note that any node a is adjacent to the (new) node $\{i, j\}$ in $g \oslash (i, j)$ if and only if a is adjacent to i or j in the original network g. In other words, all links, other than those incident to neither i nor j, are links of $g \oslash (i, j)$ if and only if they are links of g. Then $\hat{g}$, the contraction of network g, is the (uniquely defined) network obtained from g by sequences of link contractions for all links in $\mathcal{E}_P$ [13].

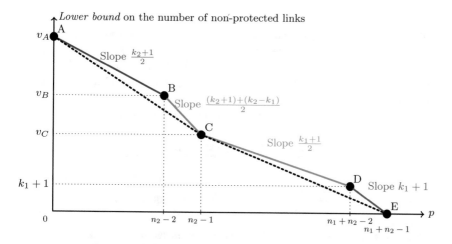

Fig. 5.2 Lower bound on the number of non-protected links as a function on the number of protected links in the IoT network. All the slopes of lines are quantified in their absolute value sense for convenience. $v_A = \frac{n_1(k_1+1)+n_2(k_2+1)}{2}$, $v_B = \frac{n_1(k_1+1)+2(k_2+1)}{2}$ and $v_C = \frac{(n_1+1)(k_1+1)}{2}$

Fig. 5.3 Illustration of network contraction. The protected links (1, 2) and (3, 4) in network g are contracted in network $\hat{g}$

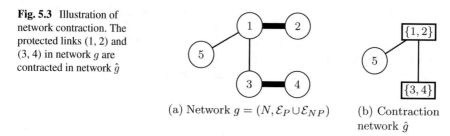

(a) Network $g = (N, \mathcal{E}_P \cup \mathcal{E}_{NP})$ (b) Contraction network $\hat{g}$

For clarity, we illustrate the contraction of a network g in Fig. 5.3. This example consists of 5 nodes and 2 protected links (represented in bold lines between nodes 1 and 2 and between nodes 3 and 4). The link $(1, 2)$ is contracted and thus both nodes 1 and 2 in g are merged into a single node denoted by $\{1, 2\}$ in $\hat{g}$. Similarly the link $(3, 4)$ is contracted. The resulting network thus consists of node 5 and supernodes $\{1, 2\}$ and $\{3, 4\}$. Since g contains a link between nodes 5 and 1 in g, then nodes 5 and $\{1, 2\}$ are connected through a link in network $\hat{g}$. Similarly, since nodes 1 and 3 are adjacent in g, then supernodes $\{1, 2\}$ and $\{3, 4\}$ are adjacent in network $\hat{g}$.

Based on network contraction, we present the proof of Proposition 5.1 as follows.

Proof Consider an IoT network g including p protected links, and $\hat{g}$ as its contraction. Let

(1) ν_1 be the number of nodes of type 1 in $\hat{g}$ (and supernodes containing only nodes of type 1),

(2) ν_2 be the number of nodes of type 2 in $\hat{g}$ (and supernodes containing only nodes of type 2),

(3) ν_0 be the number of supernodes in $\hat{g}$ that contains nodes of both type 1 and 2.

Note that if $\nu_1 + \nu_2 + \nu_0 = 1$, (i.e., if there is a unique supernode containing all nodes of the network), then no non-protected link is needed to ensure any level of (k_1, k_2)-resistancy. Otherwise, for the IoT network to be (k_1, k_2)-resistant, each element of ν_1, ν_2 and ν_0 must have a degree of (at least) $k_1 + 1$. Further, if there exist more than one element not in ν_1; i.e., if $\nu_0 + \nu_2 \geq 2$, then each of them should have a degree of (at least) $k_2 + 1$.

Thus, a lower bound on the number of non-protected links in $\tilde{s}_p^D$ is

$$\Phi = \begin{cases} \dfrac{\nu_1(k_1 + 1) + (\nu_0 + \nu_2)(k_2 + 1)}{2}, & \text{if } \nu_2 + \nu_0 > 1, \\ 0, & \text{if } \nu_1 + \nu_2 + \nu_0 = 1, \\ \dfrac{(\nu_1 + 1)(k_1 + 1)}{2}, & \text{if } \nu_1 \geq 1 \text{ and } \nu_2 + \nu_0 = 1. \end{cases}$$

Next, we focus on the study of parameters ν_0, ν_1 and ν_2. If no protected link is used, i.e., $p = 0$, then $\nu_1 = n_1$, $\nu_2 = n_2$ and $\nu_0 = 0$ and $\nu_0 + \nu_1 + \nu_2 = n_1 + n_2 = n$. Adding any protection allows to decrease the total number of elements $\nu_1 + \nu_2 + \nu_0$ by 1 (or to remain constant if the link induce a loop in a protected component of g). Thus $\nu_0 + \nu_1 + \nu_2 \geq n - p$. Similarly, for each subnetwork, we have $\nu_0 + \nu_1 \geq n_1 - p$ and $\nu_0 + \nu_2 \geq n_2 - p$. Further, the number of elements of ν_1 and ν_2 are upper bounded by the number of nodes of type 1 n_1 and type 2 n_2, respectively, i.e., $\nu_1 \leq n_1$ and $\nu_2 \leq n_2$. Finally, since $n_1 \geq 1$ then $\nu_1 + \nu_0 \geq 1$, and since $n_2 \geq 1$ then $\nu_2 + \nu_0 \geq 1$. Thus, for any p, a lower bound on the number of non-protected links in $\tilde{s}_p^D$ can be obtained by solving the following optimization problem:

$$\begin{aligned} \min_{\nu_1, \nu_2, \nu_0} \quad & \Phi \\ \text{s.t.} \quad & \nu_0 + \nu_1 + \nu_2 \geq n - p, \\ & \nu_0 + \nu_1 \geq n_1 - p, \ \nu_0 + \nu_2 \geq n_2 - p, \\ & \nu_1 \leq n_1, \ \nu_2 \leq n_2, \\ & \nu_1 + \nu_0 \geq 1, \ \nu_2 + \nu_0 \geq 1. \end{aligned} \tag{5.1}$$

To solve this optimization problem, we consider three cases.

Case 1: First, assume that $p < n_2 - 1$. From $\nu_0 + \nu_1 + \nu_2 \geq n - p$, we obtain that $\nu_0 + \nu_2 > 1$. Thus, (5.1) reduces to $\min_{\nu_1, \nu_2, \nu_0} \frac{\nu_1(k_1+1)+(\nu_0+\nu_2)(k_2+1)}{2}$ with the same constraints as in (5.1) except $\nu_0 + \nu_2 > 1$.

Since $k_2 \geq k_1$, then the minimum of the objective is obtained when $\nu_0 + \nu_2$ is minimized, i.e., when all protections involve nodes of type 2. Then, $\nu_0 + \nu_2 = n_2 - p$. Thus, the lower bound is equal to $\frac{n_1(k_1+1)+(n_2-p)(k_2+1)}{2}$. This result is illustrated by the line joining points A and B in Fig. 5.2.

Case 2: Assume that $n_2 - 1 \leq p \leq n_1 + n_2 - 2$. Then $n - p \leq n_1 + 1$. Therefore, for a given p, i.e., for a given minimal value of $\nu_0 + \nu_1 + \nu_2$, we can have either $\nu_0 + \nu_2 > 1$ or $\nu_0 + \nu_2 = 1$. Then, the lower bound of the number of non-protected links is $\min \left\{ \frac{n_1(k_1+1)+(n_2-p)(k_2+1)}{2}, \frac{(n-p)(k_1+1)}{2} \right\}$. Recall that $k_2 \geq k_1$, and therefore the lower bound achieves at $\frac{(n-p)(k_1+1)}{2}$. This observation is illustrated by the line in Fig. 5.2 joining points C and D.

Case 3: Finally, when $p = n - 1$, $\nu_0 + \nu_1 + \nu_2 = 1$, and thus no non-protected link is needed, which is represented by point E in Fig. 5.2.   □

Based on Proposition 5.1, we further comment on the locations where protected and non-protected links are placed in the two-layer IoT networks.

Corollary 5.1 *When* $0 \leq p \leq n_2 - 2$, *the protected links purely exist in subnetwork 2. When* $n_2 - 1 \leq p \leq n_1 + n_2 - 2$, *subnetwork 2 only contains protected links, and non-protected links appear in subnetwork 1 or between two layers. When* $p = n_1 + n_2 - 1$, *then all nodes in the two-layer IoT network are connected with protected links.*

Corollary 5.1 has a natural interpretation that the protected link resources are prior to be allocated to a subnetwork facing higher cyber threats, i.e., subnetwork 2 in our setting.

5.2.2.2 Networks with Special Values of p Protected Links

In the previous Sect. 5.2.2.1, we have studied for each potential number of protected links p, a lower bound $m(p)$ on the minimum number of non-protected links for an IoT network with sets of nodes $\mathcal{S}_1$ and $\mathcal{S}_2$ being (k_1, k_2)-resistant. Then, the cost associated with such networks is

$$C(p, m(p)) = p c_P + m(p) c_{NP},$$

where $C : \mathbb{N} \times \mathbb{N} \to \mathbb{R}_+$. Since the goal of the designer is to minimize its cost, we need to investigate the value of p minimizing such function $C(p, m(p))$.

In Fig. 5.2, we note that the plot of a network of equal cost (*iso-cost*) K is a line of equation $\frac{K - p c_P}{c_{NP}}$. It is thus a line of (negative) slope c_P/c_{NP} that crosses the y-axis at point K/c_{NP}. Recall also that the graph that shows $m(p)$ as a function of p is on the upper-right quadrant of its lower bound. Thus, the optimal value of p corresponds to the point where an iso-cost line meets the graph $m(p)$ for the minimal value K. From the shape of the lower bound drawn in Fig. 5.2, the points A, C and E are selected candidates leading to the optimal network construction cost. We thus investigate in the following the condition under which the lower bounds are reached at these critical points as well as the corresponding configuration of the optimal two-layer IoT networks.

Remark 5.2 Denote by s_p^D a (k_1, k_2)-resistant IoT network with p protected links and the *minimum* number of non-protected links.

Before presenting the result, we first present the definition of Harary network in the following. Recall that for a network containing n nodes being resistant to k link attacks, one necessary condition is that each node should have a degree of at least

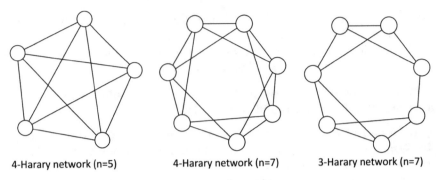

4-Harary network (n=5)   4-Harary network (n=7)   3-Harary network (n=7)

Fig. 5.4 Illustration of Harary networks with different number of nodes and security levels

$k + 1$, yielding the total number of links more than $\left\lceil \frac{(k+1)n}{2} \right\rceil$. Here, $\lceil \cdot \rceil$ denotes the ceiling operator. Harary network below can achieve this bound.

Definition 5.1 (*Harary Network* [45]) In a network containing n nodes, Harary network is the optimal design that uses the minimum number of links equaling $\left\lceil \frac{(k+1)n}{2} \right\rceil$ for the network still being connected after removing any k links.

The constructive method of general Harary network can be described with cycles as follows. It first creates the links between node i and node j such that $(|i - j| \mod n) = 1$, and then $(|i - j| \mod n) = 2$, etc. When the number of nodes is odd, then the last cycle of link creation is slightly different since $\frac{(k+1)n}{2}$ is not an integer. However, the bound $\left\lceil \frac{(k+1)n}{2} \right\rceil$ can be still be achieved. For clarity, we illustrate three cases in Fig. 5.4 with $n = 5, 7$ under different security levels $k = 2, 3$. Since Harary network achieves the bound $\left\lceil \frac{(k+1)n}{2} \right\rceil$, its computational cost of the construction is linear in both the number of nodes n and the security level k.

Then, we obtain the following result.

Proposition 5.2 *For the number of protected links p taking values of $n - 1$, $n_2 - 1$, and 0, we successively have:*

(i) each s_{n-1}^D contains exactly 0 non-protected link.

(ii) each $s_{n_2-1}^D$ contains exactly $\left\lceil \frac{(n_1+1)(k_1+1)}{2} \right\rceil$ non-protected links if and only if $k_1 + 1 \leq n_1$.

(iii) if we have the following assumptions: (i) $k_1 \mod 2 = 1$, where mod denotes the modulus operator, (ii) $n_2 > k_2 - k_1$ and (iii) $n_2 \frac{k_1+1}{2} \leq n_1$, then each s_0^D contains exactly $\left\lceil \dfrac{n_1(k_1 + 1) + n_2(k_2 + 1)}{2} \right\rceil$ non-protected links.

Proof We successively prove the three items in the proposition in the following.

(i) Note that s_{n-1}^D contains exactly $p = n - 1$ protected links. It is thus possible to construct a tree network among the set $S_1 \cup S_2$ of nodes that consists of only protected links. Thus, no non-protected link is required, and the lower bound (point E in Fig. 5.2) can be reached.

(ii) Suppose that $p = n_2 - 1$. If $k_1 + 1 \leq n_1$, we can construct any tree protected network on the nodes of S_2. Further, construct a $(k_1 + 1)$-Harary network on the nodes of $S_1 \cup \{n_1 + 1\}$, that is the nodes of type 1 and one node of type 2. Such construction is possible since $k_1 + 2 \leq n_1 + 1$. The total number of non-protected links is then exactly $\left\lceil \frac{(n_1+1)(k_1+1)}{2} \right\rceil$ (point C in Fig. 5.2). Therefore, each node in $S_1 \cup \{n_1 + 1\}$ is connected to $k_1 + 1$ other nodes, and the IoT network cannot be disconnected after removing k_1 non-protected links. In addition, the subnetwork 2 is resistant to any number of attack since it is constructed using all protected links. Note that the constructed Harary network here is optimal, in the sense that its configuration uses the least number of links for the IoT network being (k_1, k_2)-resistant.

Next, if $k_1 + 1 > n_1$, then suppose that a network g achieves the lower bound $\left\lceil \frac{(n_1+1)(k_1+1)}{2} \right\rceil$. Consider its associated contracted network $\hat{g}$. Since g contains $n_2 - 1$ protected links, then $\hat{g}$ is such that $\nu_0 + \nu_1 + \nu_2 \geq n_1 + 1$. From the shape of the lower bound Φ in the proof of Proposition 5.1, then necessarily $\nu_0 + \nu_2 = 1$ and $\nu_1 = n_1$. Thus, all nodes in S_2 need to be connected together by protected links. Since $|S_2| = n_2$, then it requires at least $n_2 - 1$ protected links, which equals p. Thus, there cannot be any protected link involving nodes in set S_1. In addition, each node in S_1 needs to be connected to at least $k_1 + 1$ other nodes in the IoT network. Since $k_1 + 1 > n_1$, then every node in S_1 should connect to at least $(k_1 + 1) - (n_1 - 1) \geq 2$ number of nodes in S_2. Recall that in a complete network of m nodes, each node has a degree of $m - 1$, and the total number of links is $\frac{m(m-1)}{2}$. Hence, our IoT network admits a completed graph in S_1 with some extra $n_1((k_1 + 1) - (n_1 - 1))$ non-protected links between two subnetworks, and in total at least $\frac{n_1(n_1-1)}{2} + n_1((k_1 + 1) - (n_1 - 1)) = n_1(k_1 + 1) - \frac{n_1(n_1-1)}{2}$ non-protected links. Then, comparing with the lower bound, the extra number of links required is $n_1(k_1 + 1) - \frac{n_1(n_1-1)}{2} - \frac{(n_1+1)(k_1+1)}{2} = \frac{n_1-1}{2}(k_1 + 1 - n_1) > 0$. Thus, $s_{n_2-1}^D$ does not achieve the lower bound (point C in Fig. 5.2) when $k_1 + 1 > n_1$.

(iii) Finally, suppose that $p = 0$. We renumber the nodes in the network according to the following sequence: $1, 2, \ldots, \frac{k_1+1}{2}, n_2, \frac{k_1+1}{2} + 1, \ldots, k_1 + 1, n_2 + 1, k_1 + 2, \ldots, 3\frac{k_1+1}{2}, n_2 + 2, \ldots$. Intuitively, we interpose one node in S_2 after every $\frac{k_1+1}{2}$ nodes in S_1. Then, we first build a $(k_1 + 1)$-Harary network among all the nodes in S_1 and S_2. Note that since $n_2 \frac{k_2+1}{2} \leq n_1$, then the last $\frac{k_1+1}{2}$ indices of the sequence only contain nodes of type 1. Thus, by construction, there are no links between any two nodes in S_2. Then, we can further construct a $(k_2 - k_1)$-Harary network on the nodes in S_2, which is possible since $n_2 > k_2 - k_1$. Thus, the constructed IoT network is (k_1, k_2)-resistant, and it is also optimal since it uses the minimum number of non-protected links. □

Proposition 5.2 and Fig. 5.2 indicate that depending on the system parameters (k_1, k_2, n_1, n_2) and for a given budget, the optimal IoT network can achieve at either point A, C or E with $p = 0, n_2 - 1, n - 1$ protected links, respectively. Notice that when $k_1 + 1 > n_1$, $s_{n_2-1}^D$ is not optimal at point C and the lower bound on the number of non-protected links is not attained. Instead, in this case, $s_{n_2-1}^D$ requires $\frac{n_1(2k_1-n_1+3)}{2}$ non-protected links in which $n_1(k_1 - n_1 + 2)$ are allocated between two subnetworks, introducing protection redundancy for nodes in S_2. For the IoT network containing 0 protected link, it reaches the lower bound (point A) if we can construct a $(k_1 + 1)$-Harary network for all nodes and an additional $(k_2 - k_1)$-Harary network for nodes only in S_2. As mentioned before, the Harary network admits an optimal configuration with the maximum connectivity given a number of links [45].

5.2.2.3 Optimal Strategy and Construction of IoT Networks

We investigate the optimal strategy and the corresponding construction for the IoT network designer in this section.

Optimal Strategy

Before presenting the main result, we comment on the scenarios that we aim to study regarding the IoT networks.

(1) First, the number of nodes is relatively large comparing with the link failure risks, i.e., $n_1 \geq k_1 + 1$ and $n_2 \geq k_2 - k_1 + 1$. Indeed, these two conditions indicate that the designer can create a secure two-layer IoT network solely using non-protected links.

(2) We further have the condition $n_2 \frac{k_1+1}{2} \leq n_1$, indicating that the type 2 nodes with higher criticality levels in S_2 constitute a relatively small portion in the IoT network comparing with these in S_1. This condition also aligns with the practice that the attacker has preferences on the nodes to compromise in the IoT which generally only contain a small subset of the entire network.

(3) Finally, we have constraints $k_1 \mod 2 = 1$ and $n_2(k_2 + 1) \mod 2 = 0$ which are only used to simplify the presentation of the chapter (whether the number of nodes and attacks is odd or even). However, they do not affect the results significantly. Note that different cases corresponding to $k_1 \mod 2 = 0$ or $n_2(k_2 + 1) \mod 2 = 1$ can be studied in a similar fashion as in our current context. The only difference is that for certain system parameters, s_0^D is not an optimal strategy comparing with $s_{n_2-1}^D$ by following a similar analysis in [13].

Therefore, based on the above conditions, the scenarios that we analyze are quite general and conform with the situations in the adversarial IoT networks. Based on Proposition 5.2, we then obtain the following result on the optimal design of secure two-layer IoT networks. Note that the solution in Proposition 5.3 is optimal to the original optimization problem presented in Sect. 5.2.1 under the considered scenarios.

Proposition 5.3 *Under the conditions that $n_1 \geq k_1 + 1, n_2 \geq k_2 - k_1 + 1, n_2 \frac{k_1+1}{2} \leq n_1, k_1 \mod 2 = 1$ and $n_2(k_2 + 1) \mod 2 = 0$, we have the following results:*

1) Regime I: if $1 + k_1 - n(k_2 - k_1) \leq 0$, then:

   (1) if $2\frac{c_P}{c_{NP}} \geq k_2 + 1 + \frac{k_2-k_1}{n_2-1}$, then s_0^D are optimal strategies.

(2) if $k_1 + 1 + \frac{k_1+1}{n_1} \leq 2\frac{c_P}{c_{NP}} < k_2 + 1 + \frac{k_2-k_1}{n_2-1}$, then $s_{n_2-1}^D$ are optimal strategies.

(3) if $2\frac{c_P}{c_{NP}} < k_1 + 1 + \frac{k_1+1}{n_1}$, then s_{n-1}^D are optimal strategies.

II) *Regime II: if* $1 + k_1 - n(k_2 - k_1) > 0$, *then:*

   (1) when $k_2 - k_1 + 1 \leq n_2 < \frac{1+k_1}{1+k_1-n_1(k_2-k_1)}$, the optimal IoT network design strategies are the same as those in regime I.

   (2) otherwise, i.e., $n_2 \geq \frac{1+k_1}{1+k_1-n_1(k_2-k_1)}$, we obtain

      (i) if $2\frac{c_P}{c_{NP}} \geq \frac{n_1(k_1+1)+n_2(k_2+1)}{n_1+n_2-1}$, then s_0^D are optimal strategies.

      (ii) if $2\frac{c_P}{c_{NP}} < \frac{n_1(k_1+1)+n_2(k_2+1)}{n_1+n_2-1}$, then s_{n-1}^D are optimal strategies.

   Thus, $s_{n_2-1}^D$ cannot be optimal in this scenario.

Proof From Proposition 5.2 and under the assumptions in the current proposition, s_0^D, $s_{n_2-1}^D$ and s_{n-1}^D achieve the lower bounds of the number of links for the network being (k_1, k_2)-resistant. In Fig. 5.2, note that the slope of the line between points A and C is $\frac{1}{2}(k_2 + 1 + \frac{k_2-k_1}{n_2-1})$, and between points C and E is $\frac{1}{2}(k_1 + 1 + \frac{k_1+1}{n_1})$, where we quantify the slopes in their absolute value sense.

In regime I, i.e., $1 + k_1 - n(k_2 - k_1) \leq 0$, we obtain $(k_2 + \frac{k_2-k_1}{n_2-1}) - (k_1 + \frac{k_1+1}{n_1}) \leq 0$, yielding that the line connecting points A and C has a higher slope than the one joining points C and E. Thus, if the lines of iso-costs have a slope higher than the slope of the line A–C, then the minimum cost is obtained at point A. Similarly, if the slope is less than that of line C–E, then the minimum cost is obtained at point E. Otherwise, the minimum is obtained at point C. Recall that the slope of the lines of iso-costs is equal to c_P/c_{NP} which leading to the result.

In the other regime II, i.e., $1 + k_1 - n(k_2 - k_1) > 0$, the slope of line A–C is not always greater than that of line C–E. Specifically, we obtain a threshold $n_2 = \frac{1+k_1}{1+k_1-n_1(k_2-k_1)}$ over which the slop of line C–E is greater than line A–C. Therefore, if $n_2 < \frac{1+k_1}{1+k_1-n_1(k_2-k_1)}$, the optimal network design is the same as those in regime I. In addition, when $n_2 \geq \frac{1+k_1}{1+k_1-n_1(k_2-k_1)}$, and if the slop of iso-costs lines, i.e., c_P/c_{NP}, is larger than the slope of the line connecting points A and E, the minimum cost is achieved at point A. Otherwise, if c_P/c_{NP} is smaller than the slop of line A–E, the optimal network configuration is obtained at point E. $\qquad\square$

From Proposition 5.3, we can conclude that in regime I, i.e., $1 + k_1 - n(k_2 - k_1) \leq 0$, when the unit cost of protected links is relatively larger than the non-protected ones, then the secure IoT networks admit an s_0^D strategy using all non-protected links. In comparison, the secure IoT networks are constructed with solely protected links when the cost per protected link is relatively small satisfying $c_P < (k_1 + 1 + \frac{k_1+1}{n_1})c_{NP}/2$. Note that the optimal network design strategy in this regime can be achieved by protecting the minimum spanning tree for a connected network. Equivalently speaking, finding a spanning tree method provides an algorithmic approach to construct the optimal network in this regime. Finally, when the cost per protected link is intermediate, the network designer allocates $n_2 - 1$ protected links connecting those critical nodes in set $\mathcal{S}_2$ while uses non-protected links to connect the nodes in $\mathcal{S}_1$. In addition, the intra-links between two subnetworks are non-protected ones.

Note that the specific configuration of the optimal IoT network is not unique according to Proposition 5.3. To enhance the system reliability and efficiency, the network designer can choose the one among all the optimal topology that minimizes the communication distance between devices.

Since the cyber threat in subnetwork 2 is more severe than that in subnetwork 1, i.e., $k_2 \geq k_1$, thus the condition of regime II in Proposition 5.3 $(1 + k_1 - n(k_2 - k_1) > 0)$ is not generally satisfied. We further have the following Corollary refining the result of optimal IoT network design in regime II.

Corollary 5.2 *Only when two subnetworks facing the same level of cyber threats, i.e., $k_1 = k_2$, the optimal IoT network design follows the strategies in regime II. Moreover, $s_{n_2-1}^D$ cannot be an optimal network design in regime II.*

Proof Based on the condition $n_1 \geq k_1 + 1$, we obtain $1 + k_1 - (n_1 + n_2)(k_2 - k_1) \leq n_1 - (n_1 + n_2)(k_2 - k_1)$. Thus, when $k_2 > k_1$, the condition of regime II $(1 + k_1 - n(k_2 - k_1) > 0)$ cannot be satisfied. Since $k_2 \geq k_1$, then only $k_1 = k_2$ yields $1 + k_1 > 0$. Therefore, $n_2 \geq \frac{1+k_1}{1+k_1 - n_1(k_2 - k_1)} = 1$ always holds which leads to the result. $\square$

Corollary 5.2 indicates that in regime II, $s_{n_2-1}^D$ cannot be optimal and the designer constructs the multi-layer IoT network using either all protected links or all non-protected links. This fact is consistent with the homogeneous security requirements since we can view the two-layer IoT networks as a unified one in this scenario, i.e., $k_1 = k_2$. Thus, the optimal design strategy achieves at boundaries either of s_0^D or s_{n-1}^D. We then simplify the conditions leading to regime I and II as follows.

Corollary 5.3 *The IoT network design can be divided into two regimes according to the cyber threat levels. Specifically, when $k_2 > k_1$, the optimal design strategy follows the one in regime I in Proposition 5.3, and otherwise $(k_1 = k_2)$ follows the one in regime II.*

Note that Corollary 5.3 presents a simpler condition than the one in Proposition 5.3 for determining which regime the optimal IoT network design lies in. We illustrate the optimal design strategies in Fig. 5.5 according to the heterogeneous security requirements and link creation costs ratio.

Robust Optimal Strategy

One interesting phenomenon is that some strategies are optimal for a class of security requirements. Thus, these strategies are robust in spite of the dynamics of cyber threat levels. We summarize the results in the following Corollary.

Corollary 5.4 *Consider to design a (k_1, k_2)-resistant IoT network. If s_{n-1}^D is the optimal strategy, then it is robust and optimal to security requirement for the network being (k_1', k_2')-resistant, for all $k_1' > k_1$ and all $k_2' > k_2$. If $s_{n_2-1}^D$ is the optimal strategy, then it is robust and optimal to cyber threat levels (k_1, k_2'), for all $k_2' > k_2$. Furthermore, the optimal strategy s_0^D is not robust to any other security standards (k_1', k_2'), for $k_1' \neq k_1$ and $k_2' \neq k_2$.*

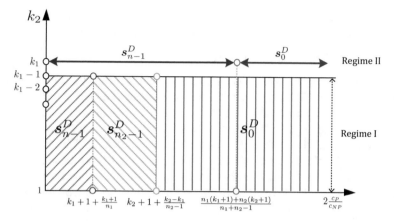

Fig. 5.5 Optimal design of two-layer IoT networks in two regimes in terms of system parameters. When $k_2 > k_1$, the optimal network design follows from the strategies in regime I which can be in any s^D_{n-1}, $s^D_{n_2-1}$ or s^D_0 depending on the value of $\frac{c_P}{c_{NP}}$. When $k_2 = k_1$, the IoT network designer chooses strategies from regime II, either of s^D_{n-1} or s^D_0 in term of the link cost ratio $\frac{c_P}{c_{NP}}$

Corollary 5.4 has a natural understanding on the selection of robust strategies. When the cyber threat level increases, then the optimal network s^D_{n-1} remains to be optimal since the network construction cost does not increase under s^D_{n-1}. Under the optimal $s^D_{n_2-1}$, subnetwork 2 is connected with all protected links and the rest is connected by a Harary network with the minimum cost. If subnetwork 2 faces more attacks, (k_2 becomes larger), then $s^D_{n_2-1}$ is robust and optimal in the sense that subnetwork 2 remains secure and no other non-protected link is required.

Robust strategies are crucial in the scenarios that the cyber threats are not perfectly perceived or they change dynamically due to the uncertain behavior of the attacker. Thus, the network designer can use a robust optimal strategy to defend against a class of cyber threats. We further illustrate this finding using a case study in Sect. 5.2.3.3.

Construction of the Optimal Secure IoT Networks

We present the constructive methods of optimal IoT networks with parameters in different regimes based on Proposition 5.3.

Specifically, the optimal s^D_{n-1} can be constructed by any tree network using protected links. In addition, the optimal networks $s^D_{n_2-1}$ can be constructed in two steps as follows. First, we create a tree protected network on the nodes of $\mathcal{S}_2$. Then, we construct a $(k_1 + 1)$-Harary network on the nodes of $\mathcal{S}_1 \cup \{n_1 + 1\}$, i.e., all nodes of type 1 and one node of type 2, where a constructive method of Harary network can be found in [45].

Finally, regarding the optimal network s^D_0, we build it with the following procedure. First, we renumber the nodes according to the sequence: $1, 2, \ldots, \frac{k_1+1}{2}, n_2, \frac{k_1+1}{2} + 1, \ldots, k_1 + 1, n_2 + 1, k_1 + 2, \ldots, 3\frac{k_1+1}{2}, n_2 + 2, \ldots$ Recall that this renumbering sequence can be achieved by interpolating one node in $\mathcal{S}_2$ after every $\frac{k_1+1}{2}$ nodes in

S_1. Then, we build a $(k_1 + 1)$-Harary network among all the nodes in S_1 and S_2. Finally, we construct a $(k_2 - k_1)$-Harary network on the nodes in S_2.

Consideration of Random Link Failures

In the considered model so far, the non-protected communication link between nodes is removed with probability 1 by the attack and remains connected without attack. In general, the non-protected links face random natural failures. If we consider this random failure factor, then there is a probability that the designed optimal network will be disconnected under the joint cyber attacks and failures. We assume perfect connection of protected links and denote the random failure probability of a non-protected link by $\kappa \in [0, 1)$. Therefore, in the regime that the optimal network design is of Harary network where all links are non-protected, then under the anticipated level of cyber attacks, a single link failure of non-protected link will result in the network disconnection. Thus, the probability of network connection, i.e., mean connectivity, is equal to $(1 - \kappa)^{\left\lceil \frac{n_1(k_1+1)+n_2(k_2+1)}{2} \right\rceil - k_2} \approx (1 - \kappa)^{\frac{n_1(k_1+1)+n_2(k_2+1)-2k_2}{2}}$ which is of order $(1 - \kappa)^{\frac{n_1 k_1 + n_2 k_2}{2}}$. Similarly, under the regime that the optimal network admits $n_2 - 1$ protected links and $\left\lceil \frac{(k_1+1)(n_1+1)}{2} \right\rceil$ non-protected links, the probability of network connection under link failure is $(1 - \kappa)^{\left\lceil \frac{(k_1+1)(n_1+1)}{2} \right\rceil} \approx (1 - \kappa)^{\frac{(k_1+1)(n_1+1)}{2}}$ which is of order $(1 - \kappa)^{\frac{k_1 n_1}{2}}$. We can see that in the above two regimes, when the security requirement is not relatively high and the size of the network is not large, the current designed optimal strategy gives a relatively high mean network connectivity. In the regime that the optimal network is constructed with all protected links, then the mean network connectivity is 1 where the random failure effect is removed.

5.2.2.4 An Illustrative Example

To better understand the presented constructive methods, we develop in this section some optimal networks s_p^D for all values of p between 0 and $n_1 + n_2 - 1$ with network parameters $k_1 = 3$, $k_2 = 5$, $n_1 = 7$, and $n_2 = 3$.

Specifically, Figs. 5.6 and 5.7 shows some optimal constructions of two-layer networks. Nodes of type 1 are represented in white circles while nodes of type 2 are represented with black dots. Non-protected links are drawn in normal lines while protected links are represented in thick lines. The figures present possible configurations for p growing from 0 to $n - 1 = 9$ respectively. For each subfigure, the caption gives the number of non-protected links needed and compares it with the lower bound computed from Proposition 5.1. Note that the lower bounds are reached in the examples except for the cases where the number of nodes in the contraction network $\hat{g}$ is not sufficient to construct proper Harary networks. Recall that network contraction and its corresponding parameters can be found in Sect. 5.2.2.1. Specifically, when $p \leq n_2 - 1$ and $n_2 - p \leq k_2 - k_1$, we have $\nu_2 = k_2 - k_1 = 2$, representing the number of type 2 nodes in $\hat{g}$ as shown in Fig. 5.6b. In addition, when $p > n_2$ and

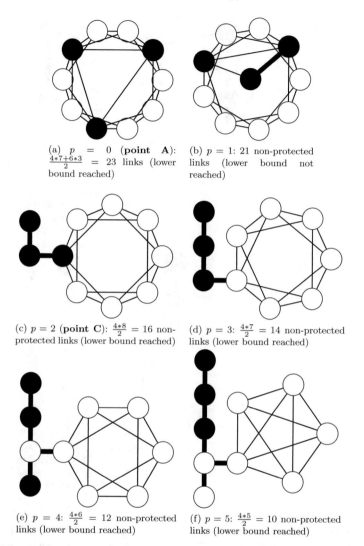

(a) $p = 0$ (**point A**): $\frac{4*7+6*3}{2} = 23$ links (lower bound reached)

(b) $p = 1$: 21 non-protected links (lower bound not reached)

(c) $p = 2$ (**point C**): $\frac{4*8}{2} = 16$ non-protected links (lower bound reached)

(d) $p = 3$: $\frac{4*7}{2} = 14$ non-protected links (lower bound reached)

(e) $p = 4$: $\frac{4*6}{2} = 12$ non-protected links (lower bound reached)

(f) $p = 5$: $\frac{4*5}{2} = 10$ non-protected links (lower bound reached)

Fig. 5.6 Optimal networks for different p under $k_1 = 3, k_2 = 5, n_1 = 7$, and $n_2 = 3$. Type 1 nodes: white circles; Type 2 notes: black dots; Non-protected links: thick lines; Protected links: thick lines

$1 < n - p \leq k_1$, we obtain $\nu_1 = 4$, 3 and 2, indicating the number of type 1 nodes in $\hat{g}$ corresponding to Figs. 5.7a–5.6c, respectively.

From Proposition 5.3, the system parameters fall in regime I. Then, we further obtain the optimal network configurations depending on the cost ratio $\frac{c_P}{c_{NP}}$ as follows.

 (i) Network depicted in Fig. 5.6a is optimal iff $\frac{c_P}{c_{NP}} \geq 3.5$,
 (ii) Network depicted in Fig. 5.6c is optimal iff $16/7 < \frac{c_P}{c_{NP}} \leq 3.5$,
 (iii) Network depicted in Fig. 5.7d is optimal iff $\frac{c_P}{c_{NP}} < 16/7$.

Fig. 5.7 (Continued):
optimal networks for
different p under $k_1 = 3$,
$k_2 = 5, n_1 = 7$, and $n_2 = 3$

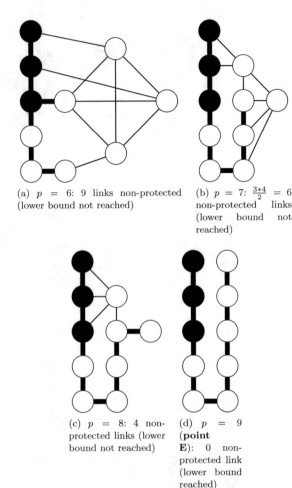

(a) $p = 6$: 9 links non-protected (lower bound not reached)

(b) $p = 7$: $\frac{3*4}{2} = 6$ non-protected links (lower bound not reached)

(c) $p = 8$: 4 non-protected links (lower bound not reached)

(d) $p = 9$ (**point E**): 0 non-protected link (lower bound reached)

Optimal IoT Network Evolution with Varying p: Figs. 5.6 and 5.7 also gives insight on the evolution of the IoT networks in a potential dynamic scenario when p evolves (due to the system constraints or change of costs). Based on the evolution of network configurations from Fig. 5.6a–d, we observe the following pattern. When a protected link needs to be removed, the optimal strategy is to remove by order of preference: (i) a protected link joining two nodes of type 1, or if no such link exists (ii) a protected link between a node of type 1 and one of type 2, and if no such link exists (iii) one protected link between two nodes of type 2. Then, the protected link that has been removed is replaced by a proper number of non-protected links for the network being resistant to adversaries. This order of removal is natural since the nodes of type 2 are more critical than type 1 nodes, and a protected link placed in subnetwork 2 can save more unprotected links.

5.2.3   Case Studies

In this section, we use case studies of IoBT to illustrate the optimal design princi-
pals of secure networks with heterogeneous components. The results in this section
are also applicable to other mission-critical IoT network applications. In a battlefield scenario, the soldiers, unmanned ground vehicles (UGV) and unmanned aerial
vehicles (UAV) execute missions together. To enhance the information transmission
quality and situational awareness of each agent in the battlefield, a secure and reliable
communication network resistant to malicious attacks is inevitable.

   The IoBT network designer determines the optimal strategy on creating links
with/without protection between agents in the battlefield. The ground layer and aerial
layer in IoBT generally face different levels of cyber threats which aim to disrupt the
network communications. Since UAVs become more powerful in the military tasks,
they are the primal targets of the attackers, and hence the UAV network faces an
increasing number of cyber threats. In the following case studies, we investigate the
scenario that the IoBT network designer anticipates more cyber attacks on the UAV
network than the soldier and UGV networks.

   To create protected D2D communication links, one method is to use moving
target defense (MTD) [43]. Specifically, instead of using a single communication
channel between agents which is easy for attackers to compromise (unprotected
link), the designer can create multiple channels and use switching strategies when
one is down. Hence, the connection of two agents through multi-channel technology
can be seen as a protected link. The cost ratio between forming a protected link and
an unprotected link $\frac{c_p}{c_{NP}}$ is critical in designing the optimal IoBT network. This ratio
depends on the number of channels used in creating a safe link though MTD. We
will analyze various cases in the following studies.

5.2.3.1   Optimal IoBT Network Design

Consider an IoBT network consisting of $n_1 = 20$ soldiers and $n_2 = 5$ UAVs ($n = 25$).
The designer aims to design the ground network and the UAV network resistant to
$k_1 = 5$ and $k_2 = 9$ attacks, respectively. Hence the global IoBT network is $(5, 9)$-
resistant. Based on Proposition 5.3, the system parameters satisfy the condition of
regime I. Further, we have two critical points $T_1 := (k_1 + 1 + \frac{k_1+1}{n_1})/2 = 3.15$ and
$T_2 := (k_2 + 1 + \frac{k_2-k_1}{n_2-1})/2 = 5.5$, at which the topology of optimal IoBT network
encounters a switching. For example, when a protected link adopts 3 channels to
prevent from attacks, i.e., $\frac{c_p}{c_{NP}} = 3$, the optimal IoBT network is an s_{24}^D graph as
shown in Fig. 5.8a. When a protected link requires 5 channels to be perfectly secure,
i.e., $\frac{c_p}{c_{NP}} = 5$, then the optimal IoBT network is of s_4^D configuration which is depicted
in Fig. 5.8b. In addition, if the cyber attacks are difficult to defend against (e.g., require
7 channels to keep a link safe, i.e., $\frac{c_p}{c_{NP}} = 7$), the optimal IoBT network becomes an
s_0^D graph as shown in Fig. 5.8c. The above three types of optimal networks indicate

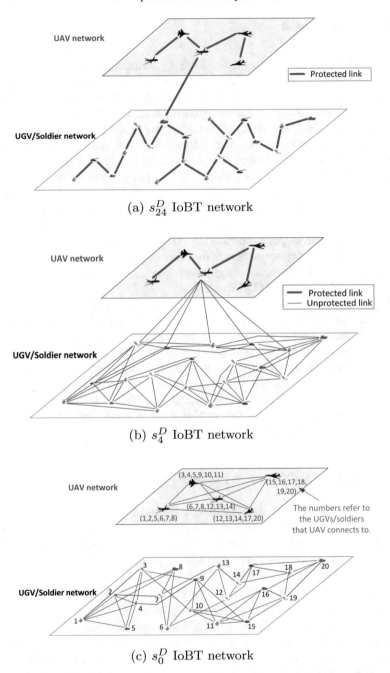

(a) s_{24}^D IoBT network

(b) s_4^D IoBT network

(c) s_0^D IoBT network

Fig. 5.8 a When $\frac{c_p}{c_{NP}} = 3 < T_1$, the optimal IoBT network is an s_{24}^D graph with all protected links. **b** When $T_1 < \frac{c_p}{c_{NP}} = 5 < T_2$, the optimal network is an s_4^D graph, where the UAV network is connected with protected links and the ground network with all unprotected links. **c** When $\frac{c_p}{c_{NP}} = 7 > T_3$, the optimal IoBT network adopts an s_0^D configuration with all unprotected links

that the smaller the cost of a protected link is, the more secure connections are formed starting from the UAV network to the ground network.

5.2.3.2 Resilience of the IoBT Network

The numbers of UAVs, UGVs and soldiers can be dynamically changing. To study the resilience of the designed network, we first investigate the scenario that a number of UGVs/soldiers join the battlefield which can be seen as army backups. As n_1 increases, the threshold T_1 decreases slightly while T_2 remains unchanged. Therefore, the optimal IoBT network keeps with a similar topology except that the newly joined UGVs/soldiers connect to a set of their neighbors. To illustrate this scenario, we present the optimal network with $n_1 = 22$ and $\frac{c_p}{c_{NP}} = 5$ in Fig. 5.9a, and all the other parameters stay the same as those in Sect. 5.2.3.1. When n_1 decreases, the network remains almost unchanged except those UGVs/soldiers losing communication links

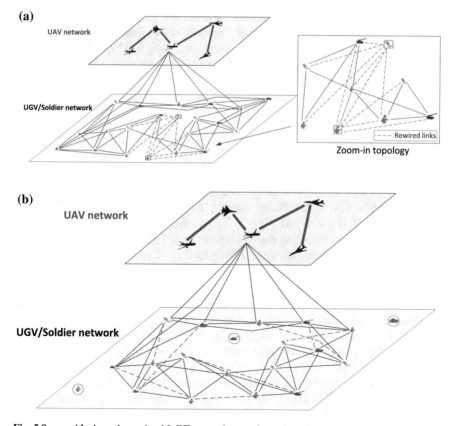

Fig. 5.9 **a** and **b** show the optimal IoBT network reconfiguration when two UGVs/soldiers join in and leave the battlefield, respectively

build up new connections with neighbors. An illustrative example with $n_1 = 17$ is depicted in Fig. 5.9b.

Another interesting scenario is that when the number of UAVs n_2 changes due to backup aerial vehicles joining in and current vehicles leaving the battlefield for maintenance. When n_2 increases, then the threshold T_1 remains the same while T_2 decreases. If the cost ratio $\frac{c_p}{c_{NP}}$ lies in the same regime with respect to T_1 and T_2 even though T_2 decreases, then under $\frac{c_p}{c_{NP}} \leq T_2$, the newly joined UAV will connect with another UAV with a protected link which either creates an S_{n-1}^D or $s_{n_2-1}^D$ graph. Otherwise, if $\frac{c_p}{c_{NP}} > T_2$, the UAV first connects to other UAVs and then connects to a set of UGVs/soldiers both with unprotected links which yields an s_0^D graph. When a number of UAVs leaving the battlefield, i.e., n_2, decreases, then T_1 stays the same and T_2 will increase under which the cost ratio $\frac{c_p}{c_{NP}}$ previous belonging to interval $\frac{c_p}{c_{NP}} \geq T_2$ may change to interval $T_1 \leq \frac{c_p}{c_{NP}} \leq T_2$. Note that regime switching can also happen when n_2 increases. Therefore, the optimal IoBT network switches from s_0^D to $s_{n_2-1}^D$ (for the increase of n_2 case, the switching is in a backward direction). For example, when the network contains $n_2 = 6$ UAVs and $\frac{c_p}{c_{NP}} = 5.4$, and the other parameters are the same as those in Sect. 5.2.3.1, from Proposition 5.3, the optimal IoBT network is an s_0^D graph. However, Fig. 5.8b shows that the optimal network adopts an s_4^D topology when $n_2 = 5$. Therefore, by adding a UAV to the aerial layer, the optimal IoBT network switches from s_4^D to s_0^D in this scenario. The interpretation is that a smaller number of UAVs is easier for the aerial network to defend against attacks, and hence protected links are used between UAVs instead of redundant unprotected links.

5.2.3.3 Flexible Design and Robust Strategies

In this section, we further investigate the secure IoBT network design in the presence of varying levels of cyber threats. Specifically, the parameters are selected as follows: $n_1 = 20$, $n_2 = 10$, $k_1 = 5$, and $\frac{c_p}{c_{NP}} = 5$. The security requirement k_2 takes a value varying from 5 to 14, modeling the dynamic or uncertain behaviors of the attacker targeting at the critical UAV network. The optimal IoBT network design is depicted in Fig. 5.10, and the corresponding cost is in shown Fig. 5.11. When $k_2 \in [\![5, 8]\!]$, the optimal IoBT network is constructed with all non-protected links. Since k_2 becomes larger, the number of non-protected links used is increasing, and thus the total cost increases. The optimal network topology switches from s_0^D to s_9^D when k_2 exceeds the threshold 8. Then, when $k_2 \in [\![9, 14]\!]$, the optimal IoBT network is unchanged as well as the associated construction cost. Despite the increases in k_2, no additional links are required since the UAV network (subnetwork 2) is connected with all protected links. Note that s_9^D is a robust strategy in the sense that the IoBT network can be $(5, k_2)$-resistant, for all $k_2 \in [\![9, 14]\!]$. This study can be generalized to the cases when the network designer has an uncertain belief on the attacker's strategy. Therefore, the IoBT designer can prepare for a number of attacking scenarios and choose from these designed strategies in the field with a timely and flexible manner.

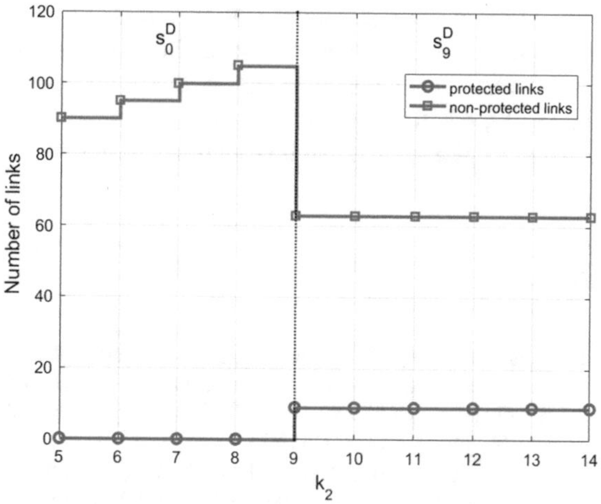

Fig. 5.10 Optimal IoBT network design with parameters $n_1 = 20$, $n_2 = 10$, $k_1 = 5$, $\frac{c_P}{c_{NP}} = 5$, and k_2 taking a value from 5 to 14. When $k_2 \in [\![5, 8]\!]$, the optimal network design is in the form of s_0^D. When $k_2 \in [\![9, 14]\!]$, the optimal network admits a strategy of s_9^D. Note that s_9^D is robust to a dynamic or varying number of cyber attacks ranging from 9 to 14

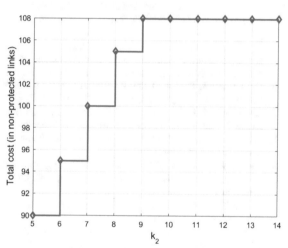

Fig. 5.11 The total cost of optimal network design in terms of the number of non-protected links. In the regime of $k_2 \in [\![5, 8]\!]$, with a larger k_2, the IoBT network requires more non-protected links to be resistant to attacks. In the regime of $k_2 \in [\![9, 14]\!]$, the total cost remains the same, since the UAV network is connected with all protected links and no additional non-protected link is required despite the increasing cyber threats k_2

5.3 Summary and Notes

In this chapter, we have studied a two-layer secure network formation problem for IoT networks in which the network designer aims to form a two-layer communication network with heterogeneous security requirements while minimizing the cost of using protected and unprotected links. We have shown a lower bound on the number of non-protected links of the optimal network and developed a method to construct networks that satisfy the heterogeneous network design specifications. We have demonstrated the design methodology in the IoBT networks. It has been shown that the optimal network can reconfigure itself adaptively as nodes enter or leave the system. In addition, the optimal IoBT network configuration may encounter a topological switching when the number of UAVs changes. We have further identified the optimal design strategies that can be robust to a set of security requirements.

The readers interested in the secure and resilient network design can refer to [6, 13, 22, 44] for more information. Further, other works on network design from economics perspective can be found in [12, 46–48].

References

1. Chen J, Zhu Q (2017) A game-theoretic framework for resilient and distributed generation control of renewable energies in microgrids. IEEE Trans Smart Grid 8(1):285–295
2. Zhu Q (2019) Multilayer cyber-physical security and resilience for smart grid. In: Smart grid control. Springer, Berlin, pp 225–239
3. Chen J, Zhu Q (2018) A stackelberg game approach for two-level distributed energy management in smart grids. IEEE Trans Smart Grid 9(6):6554–6565
4. Xu Z, Zhu Q (2015) Secure and resilient control design for cloud enabled networked control systems. In: Proceedings of the first ACM workshop on cyber-physical systems-security and/or privacy, pp 31–42
5. Dey KC, Rayamajhi A, Chowdhury M, Bhavsar P, Martin J (2016) Vehicle-to-vehicle (v2v) and vehicle-to-infrastructure (v2i) communication in a heterogeneous wireless network-performance evaluation. Transp Res Part C Emerg Technol 68:168–184
6. Chen J, Touati C, Zhu Q (2019) Optimal secure two-layer IoT network design. IEEE Trans Control Netw Syst. https://doi.org/10.1109/TCNS.2019.2906893
7. Abomhara M, Køien GM (2015) Cyber security and the internet of things: vulnerabilities, threats, intruders and attacks. J Cyber Secur 4(1):65–88
8. Lazos L, Krunz M (2011) Selective jamming/dropping insider attacks in wireless mesh networks. IEEE Netw 25(1):30–34
9. Farooq MJ, Zhu Q (2018) On the secure and reconfigurable multi-layer network design for critical information dissemination in the internet of battlefield things (iobt). IEEE Trans Wirel Commun 17(4):2618–2632
10. Lassen T (2014) Long-range rf communication: Why narrowband is the de facto standard. Texas Instruments, Technical report
11. Li Z, Zozor S, Drossier JM, Varsier N, Lampin Q (2017) 2d time-frequency interference modelling using stochastic geometry for performance evaluation in low-power wide-area networks. In: IEEE international conference on communications (ICC), pp 1–7
12. Dziubiński M, Goyal S (2013) Network design and defence. Games Econ Behav 79:30–43
13. Bravard C, Charroin L, Touati C (2017) Optimal design and defense of networks under link attacks. J Math Econ 68:62–79
14. Weber RH (2010) Internet of things-new security and privacy challenges. Comput Law Secur Rev 26(1):23–30
15. Farooq MJ, Zhu Q (2017) Cognitive connectivity resilience in multi-layer remotely deployed mobile internet of things. In: IEEE global communications conference, pp 1–6
16. Pawlick J, Chen J, Zhu Q (2019) iSTRICT: an interdependent strategic trust mechanism for the cloud-enabled internet of controlled things. IEEE Trans Inf Forensics Secur 14(6):1654–1669
17. Nia AM, Jha NK (2016) A comprehensive study of security of internet-of-things. IEEE Trans Emerg Top Comput
18. Parno B, Perrig A, Gligor V (2005) Distributed detection of node replication attacks in sensor networks. In: IEEE symposium on security and privacy, pp 49–63
19. Khouzani M, Sarkar S (2011) Maximum damage battery depletion attack in mobile sensor networks. IEEE Trans Autom Control 56(10):2358–2368

20. Vasserman EY, Hopper N (2013) Vampire attacks: draining life from wireless ad hoc sensor networks. IEEE Trans Mob Comput 12(2):318–332

21. Chen J, Zhu Q (2017) Security as a service for cloud-enabled internet of controlled things under advanced persistent threats: a contract design approach. IEEE Trans Inf Forensics Secur 12(11):2736–2750

22. Chen J, Touati C, Zhu Q (2017) A dynamic game analysis and design of infrastructure network protection and recovery. ACM SIGMETRICS Perform Eval Rev 45(2):125–128

23. Chen J, Zhu Q (2019) Interdependent strategic security risk management with bounded rationality in the internet of things. IEEE Trans Inf Forensics Secur. https://doi.org/10.1109/TIFS. 2019.2911112

24. Mukherjee A (2015) Physical-layer security in the internet of things: sensing and communication confidentiality under resource constraints. Proc IEEE 103(10):1747–1761

25. Walters JP, Liang Z, Shi W, Chaudhary V (2007) Wireless sensor network security: a survey. Secur Distrib Grid Mob Pervasive Comput 1:367

26. Zhu Q, Bushnell L (2013) Networked cyber-physical systems: Interdependence, resilience and information exchange. In: Annual Allerton conference on communication, control, and computing (Allerton), pp 763–769

27. Huang L, Chen J, Zhu Q (2017) A factored mdp approach to optimal mechanism design for resilient large-scale interdependent critical infrastructures. In: Workshop on modeling and simulation of cyber-physical energy systems (MSCPES). CPS Week, pp 1–6

28. Huang L, Chen J, Zhu Q (2017) A large-scale markov game approach to dynamic protection of interdependent infrastructure networks. In: International conference on decision and game theory for security. Springer, pp 357–376

29. Huang L, Chen J, Zhu Q (2018) Factored markov game theory for secure interdependent infrastructure networks. In: Game theory for security and risk management. Springer, pp 99–126

30. Huang L, Chen J, Zhu Q (2018) Distributed and optimal resilient planning of large-scale interdependent critical infrastructures. In: Winter simulation conference (WSC), pp 1096–1107

31. Chen J, Zhu Q (2018) Security investment under cognitive constraints: A gestalt nash equilibrium approach. In: 52nd annual conference on information sciences and systems (CISS), pp 1–6

32. Chen J, Zhou L, Zhu Q (2015) Resilient control design for wind turbines using markov jump linear system model with lévy noise. In: IEEE international conference on smart grid communications (SmartGridComm), pp 828–833

33. Chen J, Zhu Q (2017) Interdependent strategic cyber defense and robust switching control design for wind energy systems. In: IEEE power & energy society general meeting, pp 1–5

34. Chen J, Zhu Q (2016) Optimal contract design under asymmetric information for cloud-enabled internet of controlled things. In: International conference on decision and game theory for security. Springer, pp 329–348

35. Chen J, Zhu Q (2018) A linear quadratic differential game approach to dynamic contract design for systemic cyber risk management under asymmetric information. In: 2018 56th annual Allerton conference on communication, control, and computing (Allerton), IEEE, pp 575–582

36. Altman E, Singhal A, Touati C, Li J (2016) Resilience of routing in parallel link networks. In: International conference on decision and game theory for security. Springer, pp 3–17

37. Huang L, Zhu Q (2019) Adaptive strategic cyber defense for advanced persistent threats in critical infrastructure networks. ACM SIGMETRICS Perform Eval Rev 46(2):52–56

38. Chen J, Zhu Q (2016a) Interdependent network formation games with an application to critical infrastructures. In: American control conference (ACC). IEEE, pp 2870–2875

39. Chen J, Zhu Q (2016b) Resilient and decentralized control of multi-level cooperative mobile networks to maintain connectivity under adversarial environment. In: Conference on decision and control (CDC), IEEE, pp 5183–5188

40. Chen J, Touati C, Zhu Q (2019) A dynamic game approach to strategic design of secure and resilient infrastructure network. IEEE Trans Inf Forensics Secur. https://doi.org/10.1109/TIFS. 2019.2924130

41. Farooq MJ, Zhu Q (2018) A multi-layer feedback system approach to resilient connectivity of remotely deployed mobile internet of things. IEEE Trans Cogn Commun Netw 4(2):422–432
42. Gross JL, Yellen J (2004) Handbook of graph theory. CRC Press, Boca Raton
43. Zhu Q, Başar T (2013) Game-theoretic approach to feedback-driven multi-stage moving target defense. In: International conference on decision and game theory for security. Springer, pp 246–263
44. Chen J, Touati C, Zhu Q (2017) Heterogeneous multi-layer adversarial network design for the IoT-enabled infrastructures. In: IEEE global communications conference, pp 1–6
45. Harary F (1962) The maximum connectivity of a graph. Proc Natl Acad Sci 48(7):1142–1146
46. Goyal S, Vigier A (2014) Attack, defence, and contagion in networks. Rev Econ Stud 81(4):1518–1542
47. Acemoglu D, Malekian A, Ozdaglar A (2016) Network security and contagion. J Econ Theory 166:536–585
48. Hoyer B, Jaegher KD (2016) Strategic network disruption and defense. J Public Econ Theory 18(5):802–830

Chapter 6
Conclusion and Future Work

6.1 Summary

This book has investigated the resilient design and analysis of interdependent networks using game and decision theoretic approaches. To address the distinct challenges arising from interdependencies, theoretical frameworks on the network-of-networks have been established which facilitates a holistic design of interdependent networks. The book has analyzed resilient interdependent networks design across different dimensions: from static networks to dynamic networks and from finite networks to large-scale complex networks. We summarize this book as follows.

In Chap. 2, we have reviewed the basics of game theory and network science which play crucial roles in developing system frameworks and analysis in the rest of the book. In Chap. 3, we have provided a system-of-systems approach for distributed operation of multilayer networks. Specifically, we have used a game-theoretic framework to capture the uncoordinated decision making of network designers (players) where each designer controls his own layer of network. The interdependencies are reflected by the common objectives of players that maximize the integrated network connectivity. Both static and dynamic meta-network modeling have been proposed. For the dynamic MAS, the devised games-in-games framework has successfully enabled the decentralized control of agents that preserves network security and resilience. We have further provided computationally efficient methods for the agile operation of interdependent networks. In Chap. 4, we have shifted the focus from finite networks to complex networks consisting of a large population. To that end, we have established a degree-based mean field model capturing the network structure and dynamics, and studied the strategic control of two interdependent epidemics spreading over complex networks. The obtained structural results, e.g., non-coexistence phenomenon of epidemics and network equilibrium switching, have provided an optimal approach to suppressing the virus spreading. The designed quarantining strategy can be applied in a number of emerging scenarios including social network security and cybersecurity. We have further explored the secure design of

© The Author(s), under exclusive license to Springer Nature Switzerland AG 2020
J. Chen and Q. Zhu, *A Game- and Decision-Theoretic Approach to Resilient Interdependent Network Analysis and Design*, SpringerBriefs in Control, Automation and Robotics, https://doi.org/10.1007/978-3-030-23444-7_6

interdependent infrastructure network in Chap. 5. Different from the setup in Chaps. 3 and 5 has focused the network design with heterogeneous security requirement at each layer under the adversarial environment. Furthermore, the goal of the global network designer is to keep the network connected using protected and unprotected links. We have explicitly characterized the optimal strategy and provided an algorithm to construct the optimal two-layer network satisfying the requirements. The strategy has been shown with agile resilience as the number of nodes changes in the network.

6.2 Future Work

The frameworks introduced in this book would lead to many research problems in the future. In the static interdependent network formation game in Chap. 3, the link has been modeled by a binary variable. However, we can consider more general weighted links that capture the link strength between nodes as in [1]. In this way, the approximation errors resulting from mixed-integer programming can be avoided. However, additional challenge on the simultaneous link selection and weight determination needs to be addressed. As for dynamic network resilience game presented in Chap. 3, we can further consider the network operators having different estimations of severity of attacks [2], and design the multilayer MAS networks with heterogeneous security requirements. This adversarial model captures network designers' perceptions on cyber risks. Theoretically, another research direction is to design mechanisms to drive the multilayer MAS to a desired meta-equilibrium if multiple equilibria are possible. This research direction is important to enhance the network-of-networks efficiency. Some other directions include designing the multilayer MAS based on reinforcement learning and mitigating the system-of-systems security risks through strategic trust [3, 4], insurances [5], and contracts [6, 7].

The work presented in Chap. 4 has only considered two interdependent epidemics. Depending on the application scenarios, this framework can be insufficient. Thus, one future work is to extend the framework to multi-strains and derive new network equilibria and stability results. Second, we have only focused on a competing mechanism between two epidemics. The extensions to other types of interdependencies are also possible, e.g., coexistence and mutation of viruses. Third, we can investigate the epidemics quarantine under some control structures. Instead of controlling the agents in the entire degree classes which may be impossible, the system operator can only apply efforts to a subset of them which is similar to the scenarios in [8, 9]. Thus, the selection of degree classes to allocate control resources becomes critical.

In Chap. 5, the interdependent network is designed by a global operator with heterogeneous security requirements. Inspired by the model in Chap. 3, a natural next step is to extend the single network designer problem to a two-player one, where each player designs their own subnetwork in a decentralized fashion. In addition, the interdependent critical infrastructure may be composed of multiple layers, e.g., power–transportation–water triple nexus. Hence, another direction will be generalizing the current bi-level network to more than two layers and designing the optimal strategies. Furthermore, similar to [10], we can extend the current static network design to dynamic ones by considering timing of attack and recovery.

References

1. Chen J, Zhu Q (2016) Resilient and decentralized control of multi-level cooperative mobile networks to maintain connectivity under adversarial environment. In: IEEE conference on decision and control (CDC), pp 5183–5188
2. Chen J, Touati C, Zhu Q (2019) Optimal secure two-layer IoT network design. IEEE Trans Control Netw Syst. https://doi.org/10.1109/TCNS.2019.2906893
3. Pawlick J, Zhu Q (2017) Strategic trust in cloud-enabled cyber-physical systems with an application to glucose control. IEEE Trans Inf Forensics Secur 12(12):2906–2919
4. Pawlick J, Chen J, Zhu Q (2019) iSTRICT: an interdependent strategic trust mechanism for the cloud-enabled internet of controlled things. IEEE Trans Inf Forensics Secur 14(6):1654–1669
5. Zhang R, Zhu Q, Hayel Y (2017) A bi-level game approach to attack-aware cyber insurance of computer networks. IEEE J Sel Areas Commun 35(3):779–794
6. Chen J, Zhu Q (2017) Security as a service for cloud-enabled internet of controlled things under advanced persistent threats: a contract design approach. IEEE Trans Inf Forensics Secur 12(11):2736–2750
7. Chen J, Zhu Q (2018) A linear quadratic differential game approach to dynamic contract design for systemic cyber risk management under asymmetric information. In: 2018 56th annual allerton conference on communication, control, and computing (Allerton). IEEE, pp 575–582
8. Chen J, Zhu Q (2018) Security investment under cognitive constraints: a gestalt nash equilibrium approach. In: 52nd annual conference on information sciences and systems (CISS), pp 1–6
9. Chen J, Zhu Q (2019) Interdependent strategic security risk management with bounded rationality in the internet of things. IEEE Trans Inf Forensics Secur. https://doi.org/10.1109/TIFS.2019.2911112
10. Chen J, Touati C, Zhu Q (2019) A dynamic game approach to strategic design of secure and resilient infrastructure network. IEEE Trans Inf Forensics Secur. https://doi.org/10.1109/TIFS.2019.2924130

Printed in the United States
By Bookmasters